DINOSAURS
of DARKNESS

LIFE OF THE PAST

James O. Farlow, editor

DINOSAURS
of DARKNESS

Thomas H. Rich and Patricia Vickers-Rich

Indiana University Press   Bloomington and Indianapolis

FRONTISPIECE. VIEW OF DINOSAUR COVE, VICTORIA, AUSTRALIA.

THIS BOOK IS A PUBLICATION OF

INDIANA UNIVERSITY PRESS
601 NORTH MORTON STREET
BLOOMINGTON, INDIANA 47404-3797 USA

WWW.INDIANA.EDU/~IUPRESS

TELEPHONE ORDERS   800-842-6796
FAX ORDERS   812-855-7931
ORDERS BY E-MAIL   IUPORDER@INDIANA.EDU

MANUFACTURED IN THE UNITED STATES OF AMERICA

LIBRARY OF CONGRESS CATALOGING-IN-PUBLICATION DATA

RICH, THOMAS H.
    DINOSAURS OF DARKNESS / THOMAS H. RICH, PATRICIA VICKERS-RICH.
        P.   CM. — (LIFE OF THE PAST)
    INCLUDES BIBLIOGRAPHICAL REFERENCES AND INDEX.
    ISBN 0-253-33773-9 (ALK. PAPER)
        1. DINOSAURS—AUSTRALIA—VICTORIA. 2. GONDWANA (GEOLOGY)   I. RICH, PAT
    VICKERS.   II. TITLE.   III. SERIES.

    QE861.9.A82 R53 2000
    567.9'09945—DC21
                                                    00-025794

    1   2   3   4   5   05   04   03   02   01   00

To the hundreds of
volunteers and the many others,
who, over two decades, did so much
in so many different ways
to help discover the
Dinosaurs of Darkness.

He who calls what has vanished
back again into being, enjoys a bliss
like that of creating.

—Barthold Georg Niebuhr, quoted by Loren Eiseley
   in *Darwin's Century,* 1958

It is not necessary to hope
in order to persevere.

—Proverb

CONTENTS

Foreword

The past of Australia is the most mysterious of the history of all the continents. Much of its landscape is the oldest in the world. Its heartland seemed almost uninhabitable to the Europeans who came during the last two centuries, and yet it *was* inhabited—by Aborigines who had come thousands of years before and by a fauna unlike any other in the world. Where the Aborigines came from and when they came is still a subject of debate: the last twenty million years of the history of the marsupial fauna has been intensely studied and is now well understood—a mere tick of the clock in the history of this continent.

The Europeans settled mainly in the hospitable coastal areas, but almost from the beginning, hardy souls started to explore the interior. They coined the phrase "The Ghastly Blank" for the great white stretches on the map. About twenty years ago, Tom and Pat Rich expanded the term from a geographic to a paleontologic sense—this Ghastly Blank now refers to our ignorance of vertebrate evolution in Australia before about twenty million years ago—and they set out to rectify the situation. The Ghastly Blank project was designed to find the earliest ancestors of Australia's mammals and birds, and this book is a chronicle of their efforts. It is a tale of imagination, determination, and just plain hard physical labor, including one of the most difficult jobs of paleontological excavation that has ever been undertaken. But this book is more than that. It shows how scientific research really works—never in a straight line toward a predetermined goal. The title of this book demonstrates this. What do polar dinosaurs have to do with mammals and birds? Nothing—but dinosaurs were found while looking for mammals and birds, and an important trail had to be followed, because dinosaurs at polar latitudes were almost unknown and presented fascinating environmental problems.

The question of past environments led the Riches into another important aspect of paleontological research: a new discovery must be related to phenomena known elsewhere in the world. Such study has led Tom and Pat to the North Slope of Alaska and to Argentina and to ideas for future work that will surely increase our knowledge.

But what of the early birds and mammals? Sure enough, a few of them have been found—and they would not have been if it had not been for the vast amount of excavation done during the dinosaur phase of the project.

This book is a valuable record of a notable and continuing enterprise and a worthy tribute to all who participated in it.

Frank C. Whitmore, Jr.
Former Vice Chairman
Committee for Research and Exploration
National Geographic Society

Preface

The twin objectives of this book are to introduce you to a fascinating former world of polar dinosaurs that has come to light over the past two decades as well as to tell the story of how it came to be discovered.

The past world we are going to introduce you to existed on the southeast corner of mainland Australia between 100 million and 120 million years ago. Australia was then far south of where it is located today. At that time, it was joined to Antarctica, which has remained close to where it was then, while Australia has subsequently drifted far to the north at about the rate your fingernails grow. The primary focus of the story will be on the dinosaurs and other animals that lived in that polar region, what their environment was like, and how some of them adapted to this unusual place.

Intertwined with that is the story of how we have come to know what we think we know about this long-ago place that is so different from any that exists on Earth today. This part of the book is both an account of the collecting of the fossils that is central to all that we have learned and how the scientific analysis of these fossils and the environment they once lived in was carried out. We have tried to convey what it was like to execute a scientific program in the social milieu that has prevailed in Australia during the past quarter-century. Of critical importance were the individual personalities of the hundreds of people involved. They each, to a greater or lesser extent, shaped the outcome. Had other people of equal goodwill been involved in this project instead, the results would certainly have been different. Likewise, our own lives, with the particular educational backgrounds we brought to the project and our family and professional commitments, shaped the project in a myriad of ways, as did the nature of the support for it that we received, both financial and moral.

No doubt the details of how the work was done would be different elsewhere, but many of the things we encountered are widespread. We certainly did not start out with an objective in mind identical to what we have achieved. Rather, we got there by a most roundabout route. Because a major objective of this book is to let the reader gain some understanding of how the scientific process actually works, we shall take you down many of those byways, some of which led nowhere, instead of the direct route to where we finally arrived. It is not a straight-

forward story—it is full of the ambiguities of any human's life. This is because science, like all things that humans do, is an activity that bears the stamp of the real people that do it. It is not the domain of some infallible, god-like machine.

We hope, then, to convey an appreciation of not only what the dinosaurs were like in polar southeastern Australia over 100 million years ago but also how the science was done, both in the field and in the laboratory, that made it possible for this knowledge to be gained.

Tom Rich and Patricia Vickers-Rich
May 1999

Acknowledgments

Marion Anderson, Marilyn Brown, Allan Fraser, and Patti Littlefield read early drafts of this book and made many helpful suggestions regarding both content and style. We very much appreciate the assistance of Francesco Coffa, Charles de Groot, Adrian Dyer, Draga Gelt, Patricia Komarower, Peter Menzel, Steve Morton, Peter Trusler, and Mary Walters, who, in a myriad of ways, substantially contributed to this book. Kate Babbitt, James Farlow, Jane Lyle, and Robert Sloan at Indiana University Press were most enjoyable to work with in the production of the book.

DINOSAURS
of DARKNESS

1

Dinosaur Cove

"Collect butterflies." After fifty days of tunneling for fossils in the fourth season at Dinosaur Cove, that was Tom's heartfelt response to the question, "Well, what *will* you do if you give up searching for dinosaurs?"

In 1980, a small, then unnamed cove facing the Southern Ocean in Australia's southeast yielded a few bits of fossil bone. Because of this discovery, the site was soon christened Dinosaur Cove. One hundred and six million years before, an ancient stream channel, whose soft sands and muds subsequently turned to stone, had flowed through the site.

Starting in 1984, digging in earnest for fossils there had slowly brought to light a modest collection that for the first time provided a glimpse of what the dinosaur fauna of southeastern Australia had been like about 106 million years ago. Until half a dozen years before that, only one dinosaur bone had been found in all of the Australian state of Victoria and not a lot more elsewhere on the continent.

But after fifty days in the fourth year of digging at Dinosaur Cove, all known sources of dinosaurs from southeastern Australia seemed to have been totally and utterly exhausted. Future prospects of finding more seemed dim.

Background

The living mammals and birds of Australia are the most distinctive of any continent. They are clearly a reflection of Australia's isolation, and the ways in which these groups originated and then evolved on this continent have been a prime area of study and speculation for more than a century. On the basis of the fossil record of other continents, mammals are known to have originated at about the same time as dinosaurs,[1] approximately 200 million years ago. Birds are quite ancient as well, at least 150 million years old.

Primarily on the basis of their mode of reproduction, living mammals are divided into three groups: monotremes, marsupials, and placentals. Monotremes, the platypus and echidna, lay eggs. Marsupials are born at a very immature stage, then leave their mother's body and crawl to and fasten onto a nipple, which is often, but not always, located within a natural fold of skin or pouch. This embryonic neophyte remains there continuously for a period much longer than their short gestation period. Marsupials include kangaroos, wombats, koalas, and the American Opossum. The third group, placentals, are born at a more advanced stage than marsupials. They include cows, dogs, rats, bats, whales, and us.

More than half the modern terrestrial native mammals of Australia are marsupials. Marsupials are far more diverse there than on any other continent on Earth. This has long been explained by the hypothesis that marsupials reached Australia far earlier than placentals, about the end of the Mesozoic Era (the "Age of Reptiles") or at the beginning of the Cenozoic Era (the "Age of Mammals"); that is, about 65 million years ago. The route of marsupials into Australia was presumably via Antarctica from South America, the continent with the second most diverse fauna of living marsupials (and a rich fossil history of them as well). At that time the three continents lay much closer together; they had not yet been split asunder by the processes of plate tectonics which subsequently carried Australia far north of Antarctica.

Bats reached Australia by the middle of the Cenozoic Era and possibly were present there much nearer the beginning of that era. No one has yet found the remains of a single terrestrial placental of middle Cenozoic age in Australia, a time for which the fossil record there is reasonably good. About 5 million years ago, rodents reached Australia, the only terrestrial placentals to do so unassisted by humans. Unlike the marsupials, the rodents reached Australia from Asia via the Malay Archipelago.

With the discovery of a single tooth in southeastern Queensland near the town of Murgon in 1990, doubt was first cast on the picture of Australian mammalian history as the outcome of the fortuitous early arrival of the marsupials and long absence of placentals.[2] The tooth belonged to an animal named *Tingamarra porterorum,* possibly a member of a placental ungulate group, the Condylarthra; *T. porterorum* is thought to be Early Eocene in age. Both the identification of the specimen as a placental and its age as Early Eocene have been challenged.[3] One additional similar, although significantly larger, tooth has since been discovered at the Murgon site. This find, plus the discovery of jaws of other mammals at the Murgon locality, give us reason to expect that further work there will unearth specimens which will provide the evidence necessary to decide unequivocally whether or not *T. porterorum* was a placental mammal.

The Beginning

Until the 1950s, the known record of mammals in Australia was almost totally restricted to the last 1–2 million years, only the last 1 percent of their history. This was sufficient to throw considerable light on the major episode of extinction of large mammals and birds that occurred in Australia. This was part of a worldwide phenomenon, for on most other landmasses, similar episodes of mass extinction occurred either during or after the most recent Ice Age. Prior to that event, however, the Australian fossil record of these groups was virtually unknown.

That gap, the first 99 percent of their history, attracted Professor R. A. Stirton of the University of California–Berkeley to Australia in 1953 in order to locate fossil mammals and birds sufficiently old to begin to fill in this picture. With the help of the South Australian Museum and guided by a suggestion of Sir Douglas Mawson[4] to search the country east of Lake Eyre, Stirton did find the first significant collection of terrestrial mammals and birds older than 2 million years.

By the time Stirton died in 1966, the broad outline of the evolution of these two groups in Australia for the last 10 percent of their history was known. However, the first 90 percent still remained tantalizingly elusive.

How Did It Begin?

Some people's interest in a particular topic begins imperceptibly; others can pin down the moment when such an interest bursts forth. Tom's abiding interest in Mesozoic mammals, beasties that were totally unknown in Australia until 1984, can be dated almost to the hour.

Figure 1. Drawing from *All about Dinosaurs*.

For Christmas 1953 he received a copy of the book *All about Dinosaurs*[5] by Roy Chapman Andrews, who thirty years before had led the American Museum of Natural History expeditions into Mongolia. These famous expeditions discovered dinosaur eggs as well as a vast treasure trove of dinosaur skeletons. As he read Andrews's book that afternoon in 1953, Tom learned that a person who studies fossils is a paleontologist, and he decided then and there to become one. The last chapter of the book is called "Death of the Dinosaurs." Above the chapter title is an illustration showing two small mammals eating dinosaur eggs. "Those," he thought to himself, "are the interesting animals, our ancestors when the dinosaurs were alive." That idea captivated his imagination as no other ever did. He never stopped wanting to know more and more about this little-known phase of prehistory.

It was for this reason that the missing first 90 percent of the Australian mammal record tantalized Tom so much, for the Mesozoic Era is the period when the first two-thirds of mammalian prehistory occurred.

For her Ph.D. dissertation, Pat had begun to study the fossil birds that "Stirt" had collected in Australia before his untimely death. Because of this interest, we eventually immigrated to Australia, arriving in 1973. It was not our first trip to the Antipodes, however. In 1971, we had been part of an American Museum of Natural History/South Australian Museum/Queensland Museum expedition led by Richard Tedford, a former student of Stirton, searching for fossil mammals and birds.[6]

When Pat started work at Monash University and Tom received a position at the National Museum of Victoria (now Museum Victoria), our research program was directed toward throwing light on the earlier history of mammals and birds in Australia. The first step was to find sites where such fossils could be collected.

We began searching systematically for older sites, starting in the areas where Stirton and his colleagues had previously collected. We had considerable success collecting fossils from known sites and discovering new fossil localities. This refined somewhat the picture of Australian mammalian and avian evolution during the last 10 percent of their history. But the desired older fossils simply could not be found.

The unstinting help of interested and thoroughly dedicated associates, be they colleagues, students, paid assistants, parents, or volunteers, was critical to the favorable results that we did have. Time and again the volunteers were to play a vital role in our progress. One instance of this was the discovery of a fossil bone which was to turn the direction of our research from birds and mammals to dinosaurs.

John Long and Tim Flannery are cousins who as boys collected fossils from the beaches near Beaumaris, a seaside suburb of Melbourne, under the tutelage of longtime resident Colin Macrae. Their enthusiasm has carried them to positions in the Western Australian Museum and the South Australian Museum, respectively.[7] In 1978, after having assisted with our research program for a few years, they decided to work with geologist Rob Glenie to try to find more fossil bones where the one dinosaur fossil then known from Victoria had been collected at the turn of the century.

William Hamilton Ferguson was a Mines Department of Victoria geologist who was searching for coal along the coast near the town of Inverloch in 1903. He possessed an uncanny knack for finding fossils where others could locate none. This is exactly what he did on the coast at a place immediately west of a prominent rock stack called Eagle's Nest. There he discovered and collected an isolated toe bone of a carnivorous dinosaur (Fig. 2), the first dinosaur bone found in Australia to be described in a scientific paper.[8] In his paper he described a lungfish tooth from the same site. The locality of both specimens was meticulously marked on his exquisite geological map of the area (Plate 1).

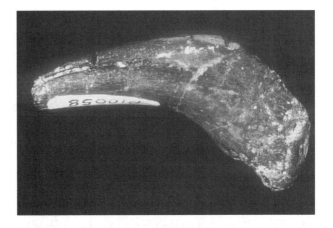

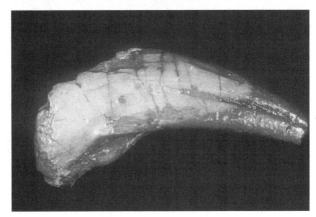

Figure 2. The toe bone of the carnivorous dinosaur found by Ferguson in 1903 near Eagle's Nest. The specimen was sent to England and described by A. Smith Woodward of what is today the Natural History Museum, London, who assigned it to the long-established English genus *Megalosaurus*. It was a frequent practice at that time for Australian fossils to be sent overseas to be scientifically analyzed, because the requisite expertise to carry out such studies was generally thought not to exist within Australia.°
Length 53 mm (2.0 inches).

°Rich (1999).

It was this map that Glenie, Long, and Flannery were following 75 years later when they returned to search for more fossils on a cold, blustery day with the sea running high. Many other paleontologists, ourselves included, had been to this area, taken a look, found nothing, and gone elsewhere. But unlike the others, almost as soon as they reached the spot indicated on Ferguson's map, John found a pebble with a bone fragment inside.

The hardest fossil to find is the first one. That is the specimen which convinces a fossil hunter that there are indeed fossils present and that more can be found.

Building on John's discovery, Tim returned time and again over the next six months to the outcrops of the same rocks that had produced this fossil between Inverloch and San Remo. By the end of that time he had collected about thirty specimens. Among them were what was obviously the femur of a small ornithischian (bird-hipped) dinosaur. But there were also two bones that defied ready identification because we had no previous, firsthand experience in analyzing fossil reptiles or amphibians (Plate 2).

Ralph Molnar of the Queensland Museum is one of the few vertebrate paleontologists in Australia with firsthand experience studying dinosaurs.[9] For his Ph.D. dissertation, he described the skull of *Tyrannosaurus rex*. Because of his encyclopedic knowledge of dinosaurs, he was soon able to make sense of one of the two enigmatic specimens, convincing himself that it was the astragalus, or ankle bone, of a dinosaur very similar to the large carnivore *Allosaurus*. This was not an expected identification, for two reasons. First, it occurred in rocks about 20–25 million years younger than those in which that dinosaur was known elsewhere. Second, it occurred in Australia. Previously, it was only known for certain to have lived in North America and possibly in East Africa.

The first scientific paper about this astragalus appeared in 1981. Two years later, Samuel P. Welles of the University of California presented arguments about why it should not be assigned to *Allosaurus*. In the leisurely manner of scientific journals, a reply to that criticism took another two years to appear, although the manuscript was written soon after the publication of Welles's paper. The final of this series of three papers made the case for assigning the astragalus to *Allosaurus*, and there the matter stood for more than a decade.[10]

But such an hypothesis is never final in science. At the present time, Daniel Chure of Dinosaur National Monument is reviewing all the fossils that have been assigned to the Allosauridae. After studying the astragalus at first hand, he is of the opinion that it is not *Allosaurus*, although probably it is a member of the broader family grouping Allosauridae. When he publishes his reasons, the matter may stand there or some other specialist may disagree and publish yet another interpretation. There is no formula for reaching a final conclusion in a matter like this. The debate could go on for decades, or when Dan states his case his arguments may be so persuasive that no further rebuttals will be made. Only time will tell. Of course, if more material of this dinosaur is discovered, it could change the story again!

The second specimen defied identification for a number of years, but its "coefficient of weirdity" alone was enough to add to the impression already provided by the ornithischian femora and the *Allosaurus* astragalus—that in Victoria, if more of the dinosaur fauna and the biota associated with it could be found, interesting things would turn up. Eventually, this specimen would be identified as a labyrinthodont amphibian[11] (Plate 2C).

In the Devonian, amphibians evolved from fish. The most prominent group among these first amphibians were the labyrinthodonts. The name refers to the structure of their teeth. When a cross-section is cut through one of them, the dentine and sometimes the enamel is complexly infolded, giving the appearance of a labyrinth—thus the name of this group. In life, these animals looked very similar to crocodilians. Just like living crocodilians, they had a sprawling posture and were carnivorous. Also like them, they were amphibious.

Labyrinthodonts flourished until the end of the Triassic. Up until the early 1980s, labyrinthodonts were thought to have become extinct at that time. But in that decade, first an Early Jurassic labyrinthodont was reported from Australia[12], then a Middle Jurassic form in China[13], and finally a Late Jurassic labyrinthodont[14] was reported from Mongolia. Then an even younger form, *Koolasuchus cleelandi,* was found in the 110-million-year-old sediments of southeastern Victoria, 30 million years after the last known form occurred anywhere else in the world. *K. cleelandi* is the youngest occurrence of this once very successful group of carnivorous lake and stream dwellers, a relict form (like today's platypus) from the Early Cretaceous of eastern Gondwana.[15]

Although it is a common conception that the living amphibians, including frogs, toads, newts, salamanders, and caecilians, are primitive terrestrial vertebrates, they, in fact, appeared relatively late in the fossil record, about the same time as the dinosaurs and mammals: the Middle to Late Triassic.

A Shift to Other Pastures

By the end of 1978, Tim Flannery had visited all the coastal outcrops of the rocks where one might reasonably expect to find fossils between

Inverloch and San Remo. As he looked at the geological map of Victoria, Tom realized that similar rocks occur in the Otway Range to the west of Melbourne.

There, too, these rocks occur extensively on the rocky shore platforms as well as inland. It was most important that some part of the rocks occur on the shore platform, because only there were fossils likely to be found. Inland, the same rocks are typically covered by farmland and forests, making it virtually impossible to find exposed, chemically unaltered rocks which might contain fossils. Even where outcrops of bare rock occur inland, by the time fossil bones there are exposed by erosion, leaching of the rock due to soil formation processes has usually dissolved the bones away. Frequently, too, the rock outcrops inland are covered with a black patina, which obscures any bones that might have survived long enough to be otherwise exposed. By contrast, on the shore platform, the mechanical process of erosion—the constant battering by the sea—outraces the soil formation processes, so that the rocks exposed are not chemically altered and are constantly swept clean, making it easier to see the fossils. If not collected soon, unfortunately, they are pounded to pieces and washed away.

Thus, although at the time no fossil bones were known to us from the Otways, we decided to prospect these rocks, which were similar to the rocks on the coast between Inverloch and San Remo, during 1979 and 1980.

Being confined to the shore platform meant that we were restricted to a tiny area to examine, a strip about 20 meters (65 feet) wide on average and 200 km (125 miles) long. That is only 4 square kilometers, or 1½ square miles. Compared to the vast regions of badlands where dinosaur bones are found in places like western North America, eastern Asia, or southern South America, this is almost nothing. Almost, but not quite—and the "not quite" is the operative phrase here. Actually, although we were confined to such a small area, we had a great stroke of luck in that the area available to us extended laterally over such a great distance. Even in the parts of the world where the richest concentrations of dinosaurs occur, they are not uniformly abundant. The Lance Formation of Wyoming is the rock unit out of which John Bell Hatcher, one of the greatest dinosaur hunters who ever lived, collected thirty-three ceratopsian skulls for Yale University in a period of four years near Lance Creek. However, if one were to search a circular area of the Lance Formation of only 4 square kilometers selected at random, it is quite possible that precious little would be found, because some areas where the formation is exposed are more likely than others to contain fossils. One can improve the odds of finding fossils if the shape of the randomly selected area is changed from a circle to a string or a long sinuous strip such as the rocks exposed on a shore platform, simply because a far greater range of rock types will likely be sampled.

On the shore platform at Eastern View, about 100 km (60 miles) southwest of Melbourne, the most northeasterly coastal outcrops of the Otway Group occur. The Otway Group is the name given to this body of rock, in which the dinosaurs from the Otway Range have been found. Eastern View was where we began our search for dinosaurs in the Otways. Walking over the shore platform, examining every available square meter of exposed rock for a sign of fossil bone, we slowly worked our way toward Cape Otway over a period of a month in our first prospecting venture into this area in 1979.

Until our peregrinations reached Marengo (just west of Apollo Bay) late in the 1979 season, days would go by without a single fleck of fossil bone being found. This was soul-destroying work, because the thoughts were always gnawing at us as we tramped endlessly, searching in every nook and cranny that could be reached, that either there were no fossils to be found and we were on a fool's errand or that we, as fools, did not recognize the bones that were there.

Finally, at Marengo, a few bone scraps turned up, but these few were enough to show that our doubts were definitely not warranted. Most of the bone fragments were no more than water-worn bits, but one was a partial toe bone of some sort of a carnivorous dinosaur. As we continued onward for a final day of prospecting, we found four sites between Blanket Bay and Cape Otway that yielded more fossils. At one site, the first dinosaur teeth from Victoria were discovered—teeth that almost smiled at the paleontologist who found them! The mammals and birds we had set out to find still completely eluded us, but at least our discoveries were encouraging enough to bring us back the next year and to continue prospecting west of Cape Otway. We were convinced there were definitely bones to be found in the hard, green-gray sandstones of the Otway Group and that, with luck, richer sites might be discovered.

West of Cape Otway, the coastal outcrops are harder to reach than those to the east. To the northeast of Apollo Bay, the Great Ocean Road hugs the coastline, which meant that it was easy to get to most of the shore platforms with a modest effort. By contrast, in 1980, we would often spend a morning finding our way down to a remote stretch of shore platform, then prospecting 100 meters (110 yards) of it before being cut off by channels which sliced through it, ending against a sheer cliff. We would climb out in time for lunch, only to repeat the process at another locality for a similar result in the afternoon.

Working slowly westward in this way, on December 15, 1980, the party entered what was later to become Dinosaur Cove. Mike Archer, then at the University of New South Wales, and Tim Flannery were

Figure 3. A cartoon quite similar to this appeared in the *Sun* newspaper at the time Dinosaur Cove was officially named. *Artist: Peter Trusler*

walking together along the base of the cliff, while Tom was about 15 meters (50 feet) away at the water's edge.[16] Mike and Tim were nattering away like a pair of magpies, and Tom was certain their attention was so diverted they could not possibly find a thing. Just about the time that thought passed through his mind, a whoop went up and they were both on their hands and knees, diligently searching the rocks.

As good fossil hunters can, even though seemingly preoccupied with other things, one of them had spotted a fragment of bone. Now crawling around, they quickly found three more. It was evident from even the briefest look that the fossils all were confined to what 106 million years ago had been the bed of a small stream channel.

Over the next few days, while part of the exploration party continued to push westward, eventually exploring outcrops beyond Moonlight Head, others dug into the sandstone in the ancient channel Mike and Tim had spotted, trying to find more. The tools were modest, and only a small quantity of rock could be recovered. Eventually, about a dozen fossils turned up, but all were merely water-worn fragments of bone which could be identified only as "vertebrate."

The name Dinosaur Cove was bestowed quite casually. The cove where this ancient former stream channel had been located needed a name. However, there was no name on the map. Every other site where fossils had been found was easy to name because of some nearby geographic feature. In this case, a name was simply invented to satisfy an immediate need as field notes were being written on the day of its discovery. Only later did it come to be seen as a most appropriate choice for what by then was an important fossil site.

In Victoria or elsewhere, one does not simply give a name to a cove (or any other geographic feature) and expect it to become generally accepted. Victoria has an official standard procedure for naming places. In the case of Dinosaur Cove, we made an initial contact with the Office of the Place Names Committee of the Department of Crown Land and Survey. We submitted the proper paperwork to make an official proposal to that body. After a time, the proposed name was officially published in the *Victorian Government Gazette* as a notice that this course of action was to be undertaken unless someone had an objection. When no objection had been raised for several months, a second notice was published in the *Victorian Government Gazette* and the name was official; it soon appeared on maps of the area (Fig. 4A). We were delighted because we had had a chance to put official names on places only twice before, once for a place in Antarctica and once for a feature on Venus!

In 1981, a year after our initial discoveries, more digging at Dinosaur Cove revealed six more bone fragments, which reinforced our impression that if more of the channel deposit could be excavated, eventually identifiable fossils would be found. However, it was obvious that

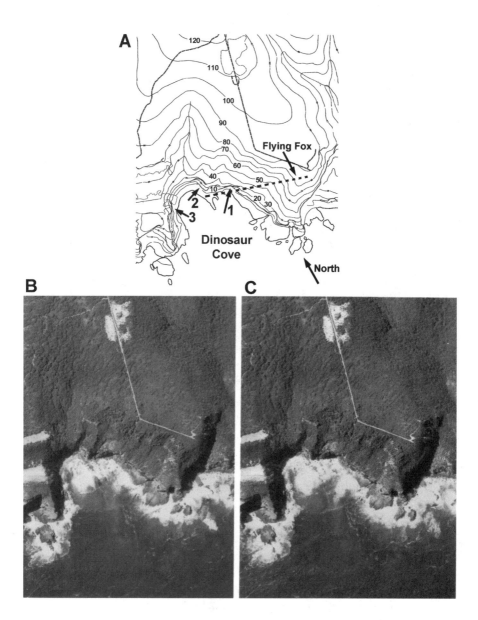

Figure 4A. Topographic map of the Dinosaur Cove area at a scale of 1:10,300. 1, Slippery Rock. 2, Dinosaur Cove East. 3, Dinosaur Cove West. The dashed line labeled "Flying Fox" shows the position of the cable for the aerial tramway. Contour interval: 10 meters (33 feet).

Figures 4B, 4C. Aerial photographs of Dinosaur Cove at the same scale as Fig. 4A. If you look at the left-hand photograph with your left eye and the right-hand one with your right eye, you will see the image in three dimensions. To make that easier to do, place a vertical piece of paper or card between your eyes so that each of them can see only one image. If you can then make the images overlap, the image will appear in three dimensions.

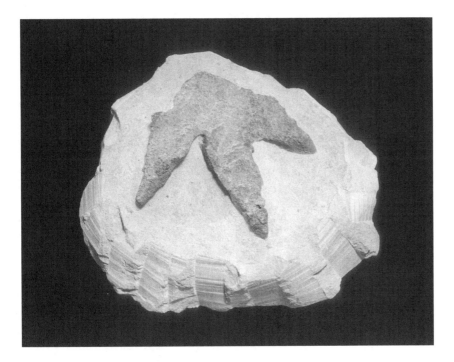

Figure 5. One of the few other discoveries in 1980, a single footprint of an ornithopod dinosaur. The reason that "footprint" stands out of the surrounding rock is that the reddish sand that filled the depression made by the dinosaur's foot became harder than the greenish sandy clay that the dinosaur stepped in. As the sea pounded the surface where the footprint was found, this difference in hardness resulted in the greenish sandy claystone eroding faster than the reddish sandstone. The site where this ichnite was found, near the mouth of the aptly named Knowledge Creek, is so difficult to reach that we have never been back.

tunneling would have to undertaken if much of this deposit was to be excavated, for the bulk of the deposit lay inside the vertical cliff face and not on the shore platform. We continued prospecting, hoping to find a site equally as rich but more accessible. None of us were, or for that matter wanted to be, underground miners.

Through an incredible set of circumstances, the National Museum of Victoria hosted an exhibition of Chinese dinosaurs in 1982. At about the same time, the Friends of the National Museum of Victoria was formed. Members of that group participated in the exhibition as volunteer guides and attendants and ran a shop that sold dinosaur paraphernalia. Unbeknown to us, this group had a growing passion to participate in a dinosaur excavation; as we will later detail, this was the trigger for the beginning of serious excavation at Dinosaur Cove.

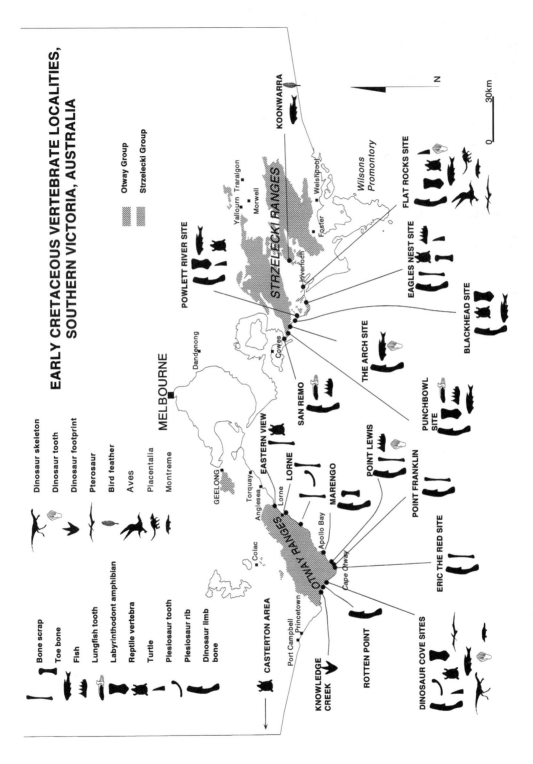

Figure 6. Map showing the principal localities where dinosaurs have been found in Victoria.

"Golden Moments"

In Harriet Beecher Stowe's book *Uncle Tom's Cabin,* there is a memorable scene where the runaway slave Eliza is fleeing her overseer, Simon Legree. Clutching her baby to her chest, she crosses the Ohio River by jumping from one ice floe to another as they drift down the river. That image can serve as a metaphor for many things. One is that a pathway in life is unique. Two researchers will never have the same sequence of events in their careers that will make their paths the same. An even more compelling interpretation is that the opportunities to jump from one ice floe to the next are ephemeral. The opportunities have to be seized at a particular moment or they are gone forever, and other paths are followed.

Many things happened along the way which were critical to creating the circumstances in which the polar dinosaur project of southeastern Australia could take place. At the time many of these events occurred, the immediate outcome one way or the other was often not seen to have much importance. But in hindsight each event was clearly crucial.

One of the first of these events was our migration to Australia. Pat had received a post-doctoral Fulbright Fellowship to spend nine months in Australia completing projects relating to her Ph.D. dissertation, which she had just submitted at Columbia University. Without employment in the United States and little prospect of it, Tom accompanied her. At that time you could come to Australia with a one-way ticket. So we needed to save enough funds during our Fulbright stay to get Tom back to the United States by the end of the fellowship. After staying in Melbourne for about one month, Tom casually purchased a copy of the *Weekend Australian* newspaper. He idly turned to the employment section, merely out of curiosity, to see what kinds of jobs were available in Australia. He had no particular expectation, or desire, to find a job for himself until our return home. The job he would eventually receive at the National Museum of Victoria was advertised on the first or second page of that section. Because the job was not advertised outside of Australia and no one had ever mentioned to him that such a position was to be advertised, the chances that such an advertisement would be placed on the day he just happened to buy the paper in a country halfway around the world from home and that he would look at a section of the paper he hardly ever read seemed at the time, and still does, incredibly unlikely. Tom got the job, and he became part of the "woodwork" at the Museum.

The first excavation at Dinosaur Cove in 1984 would never have happened if The Friends of the Museum of Victoria had not pushed Tom to carry it out. And the Friends probably would not have been in a position to push had it not been for an exhibition of Chinese dinosaurs.

About the time Dinosaur Cove was discovered, a small delegation of Chinese museum scientists came to Australia. A reception was held for them, which museum staff were invited to attend, Tom among them. In a conversation with an entomologist from the Tianjin Museum, it came up that Tom was a paleontologist. As a consequence, the entomologist pulled out a series of photographs of fossil vertebrates in his museum. The second photograph was of a skeleton of the large sauropod dinosaur *Mamenchisaurus*. In halting English, the entomologist said, "And this is going to Japan." Tom did a mental double flip. If it was going to travel to Japan, it was going to have to travel across water. If it could go across water that far, why not all the way to Australia? Why not? Tom grabbed the director of the National Museum of Victoria, Barry Wilson, by the arm and led him over to the Chinese entomologist and asked him to repeat what he had just said. Then Tom persuaded the director to write a letter to the director of the Tianjin Museum to see if it just might be possible to also loan that skeleton to the National Museum of Victoria, as they were already lending it to a Japanese newspaper company.

It just so happened that Pat was about to make her second trip to China. She had first visited there in 1979 and had gotten to know many members of the Institute of Vertebrate Paleontology and Paleoanthropology (IVPP) in Beijing, forming close friendships with several of the staff. Along with the Museum Council and the director, we decided that on this second trip she should take with her a copy of the letter to the director of the Tianjin Museum. Once in Beijing, she would ask her contacts at IVPP for some personal assistance with the loan of the *Mamenchisaurus* skeleton. A letter from a totally unfamiliar Australian museum might never be answered by the Tianjin Museum if it simply arrived unannounced and was not supported by prominent Chinese scientists. When Pat got to Beijing, she raised this matter with Minchen Chow, then director of IVPP. His response was most startling but reassuring. "Actually, that fossil belongs to IVPP, and I shall certainly help you borrow it."

As a consequence of these two equally improbable events, *Mamenchisaurus* and several other Chinese dinosaurs were displayed, first in Melbourne and then in Sydney. The *Dinosaurs from China* exhibition attracted more than 500,000 Australian visitors; it was the first dinosaur exhibition ever to tour the continent. Another outcome was a joint casting operation, which netted the Museum of Victoria casts of the skeletons of two large dinosaurs for their permanent exhibitions. Which brings us back to the Friends of the National Museum of Victoria.

From their dinosaur shop operation accompanying the *Dinosaurs*

Figure 7. Entrance to the National Museum of Victoria, Melbourne, Australia, 1982.

from China exhibition in Melbourne, the Friends made a few thousand dollars. As a result of their work with the exhibit, they became very "dinosaur conscious." They were well aware of the work that had been going on virtually in their backyard to uncover dinosaurs. That is how the Friends came to approach Tom about having a dinosaur excavation in which they could participate. Try as he might, Tom could not dissuade the Friends. Repeated explanations that there were no sites where one could go with a large party of volunteers and expect to dig up dinosaurs failed to dent their enthusiasm. Finally, Tom offered to take a group to Dinosaur Cove with the firm understanding that results were not guaranteed. They were told that a dig there might well be a complete bust and that all their effort might go for naught. They seized the opportunity with alacrity and unbridled enthusiasm. Both of us were, therefore, faced with the problems of organizing an underground excavation, a course of action we dreaded. Without that determined push from the Friends, it is quite likely that the vast bulk of Victoria's dinosaurs would still be in their underground tombs—unloved and unrecognized.

So it was that in 1984 excavations began at Dinosaur Cove.

2

The Crossing of the Rubicon

Our knowledge of tunneling in rock or in anything else at that time was completely nonexistent. Our effort to obtain expert advice was initially a matter of groping blindly. Not knowing where to begin, Tom started by talking with people at the Victorian Department of Minerals and Energy. As he was passed from one person to another, some in government departments and some in private industry, he gradually found people who did give him the professional advice he sought. Most of that initial advice was tinged with bewilderment and anxiety by those who gave it. "You are proposing to dig tunnels using amateur volunteers?" The professionals had difficulty coming to grips with this idea. And tunnel-digging volunteers are not led by people who have no previous experience themselves doing that sort of thing. Dire predictions of catastrophe of a dozen different varieties were mooted and listened to.

Talking to one professional after another gradually resulted in the identification of those individuals who could not only recognize problems but who also could come up with practical solutions. These valuable and imaginative people could see that although our proposed approach to tunneling was novel, there was no fundamental reason why, if they were suitably guided, inexperienced people could not carry out this kind of work in reasonable safety. The contributions of expert advice from these people with a positive outlook was critical, for not only did they identify the dangers that lay ahead but, equally important, they suggested practical approaches to achieving the desired results. In a few instances, Tom abandoned proposed courses of action after receiving their advice. The vitally important art of deciding whose advice to listen to was honed by this experience. Had all the naysayers held sway, excavations at Dinosaur Cove never would have taken place. On the other hand, had Tom ignored all precautionary advice, it is unlikely that the safety record of the site would be as good as it is.

Armed with this advice, we went through the exercise of drawing up a detailed plan of action, making sure that all the necessary steps were recognized and considered and that all the safety precautions were in place. The point of the exercise was to have fully thought through what needed to be done before doing it. But when the time came to do the job, we both knew it would not turn out exactly as we had foreseen and that the plan would, in the end, probably bear little relationship to the events it was drawn up to deal with. So when at long last he set out for Dinosaur Cove to begin the first tunneling effort, Tom deliberately left the plan behind at the museum and made his field decisions as the situation developed.

The initial attempt to tunnel at Dinosaur Cove was made in February 1984. To get the infrastructure in place before the main party arrived, an advance team went on site to set up a camp adjacent to a paddock (pasture) about 2 km (1¼ miles) from Dinosaur Cove. They built a cookhouse, laid telephone lines and air hoses, and positioned the compressor. Then, for sixteen days sixty-five volunteers, ranging in age from 7 to 70, struggled night and day to burrow into the cliff where the sediments of the ancient stream channel were preserved.

On Saturday the 11th of February, the full party was in camp and the Surf Life Saving Victoria helicopter arrived on schedule. The helicopter flew the heavier equipment from the paddock near the camp down to the shore platform in Dinosaur Cove. The larger timbers for the tunnel's protective entrance portico were transported this way, as were bulkier items of rock-excavating equipment. A shortage of helicopter fuel meant that many of the lighter items had to be carried in. The result was a line of people going down into the cove, each carrying a piece of equipment such as a split log, passing others climbing out to get another load. From a distance, the scene was reminiscent of a trail of ants.

Soon the gear was in place, and in an incredibly short period of time the protective portico was built. By the end of the afternoon, drilling of the rock face had begun.

For the next week, progress was slow and frustrating while the crew learned to operate the equipment and made adjustments to the equipment itself. There was no one person who knew all aspects of what was being done. The people from Atlas Copco, who had provided the equipment, had no idea about what was required to excavate fossils, and we had no idea what many of the items of equipment they provided were for. Fortunately, some of the people on the crew had a technical background that enabled them to quickly understand the bits and pieces of equipment, although none had ever worked with most of it.

Over that first week, problems, such as insufficient air pressure, leaking and bursting hydraulic lines, rock breakers (jackhammers) which would not hold their bits, and cranky electrical generators, were dealt with in one way or another. Because only sixteen days were allocated for this first dig (based both on available funds and the time volunteers were able to devote to it), it was imperative to minimize down time on the equipment. Therefore, the midnight trip of 220 km (135 miles) to Melbourne to pick up a replacement part or alternative piece of equipment became a commonplace event, as was a twenty-four-hour day on the rockface whenever possible.

As Tom tried to obtain technical assistance from Atlas Copco to solve these problems, he became frustrated about the delays. His first encounter with Bill Loads of Atlas Copco was a rather acrimonious telephone call in which Tom complained quite sharply about not getting the help needed as quickly as he thought a full-paying customer should receive. Bill calmed him down and promised to see that things were fixed as soon as possible. It is a testimony to Bill's patience that he ever spoke civilly to Tom after that first encounter, but they came to be good friends as well as close colleagues in the search for Victorian dinosaurs. Only later did we learn that Bill, who turned out to be the head of the entire Victorian operations of Atlas Copco, had ordered that support for the dig be given as a "peppercorn job." That is, the original quote was far below the market value for what was offered. In the end, because of cost overruns elsewhere in that pioneering dig, Bill saw to it that an invoice for even that amount was never presented by Atlas Copco.

Figure 8. Sign at the entrance to the camp at Dinosaur Cove.

In the years to come, because he believed that with sufficient long-term backing something exciting would be found at Dinosaur Cove, Bill remained determined to provide that kind of support. Not only did he channel resources at his own disposal but he also encouraged business associates in other industries to provide support in kind.

After the air compressor was set up there was enough pressure to drive all the equipment except the rock breakers. Air friction in the more than 200 meters (650 feet) of hose between the compressors at the top of the cliff and the tools at the bottom meant that the pressure drop was so great that those vital tools were all but useless. A few seconds of intermittent operation with a break to allow the compressor time to repressurize the hose was possible, but that was hardly adequate for an efficient operation.

One evening Tom left camp at 6:00 P.M. for Melbourne, where he was able to pick up enough rubber hose to put in a second line next to the first one. He returned to camp at 2:00 A.M., and the hose of the second airline was laid just after first light. A few minutes of testing showed that this approach was not a workable solution, so Tom had the crew pull up all the second hose line and got on the telephone again. Tom returned the hired hose for the second line quickly so that we would only have to pay for a 24-hour rental, although it meant another midnight dash to Melbourne for him. Then Tom waited for Bill Loads, who promised to come down the next day with an air receiver.

Tom was not quite sure what an air receiver was, but he had been assured that it would solve all the problems. It did. By having the airline pass into a large tank (the air receiver) near the working face and then back into a short hose that extended the remaining distance to the tool, a reservoir of high-pressure air was made available that could drive the tools.

Although several thousand dollars had been allocated for this excavation, officially Tom was only allowed to spend $200 on any one item without prior approval from the National Museum of Victoria. When trying to talk a supplier out of a vitally needed item at 3:00 A.M. on a Sunday morning, that limitation proved unworkable, so rather than let the excavation bog down, Tom accepted the prospect of an inevitable official slap on his wrist and went ahead and purchased vital items without such approval. Administrative headaches continued to dog the field operation of this project because it did not fit the well-understood, ordinary activities of the organizations that had the responsibility to manage their funds.

When the equipment was operating, the work went on around the clock. This was done in large part because the dig was restricted to sixteen days: funds for food, rental hire, and the salary of the mine supervisor would only go that far. In addition, most of the volunteers were giving up a significant part of their annual vacation to participate. Another factor was that only a few people at one time could work in the small space available. With such a large crew of volunteers, the only way to give as many people as possible an opportunity to participate in the excavating was a round-the-clock operation.

The method used to tunnel at this stage involved first drilling a series of holes in the rock. Then the rock between them was broken by splitting it with a steel wedge that slid between two pieces of steel shaped to fit the hole. The wedge was either driven by a sledgehammer or by a pneumatic-hydraulic device called a Darda. The Darda proved to be like the little girl with the little curl: when she was good, she was very, very good, but when she was bad, she was horrid. Problems with leaking and bursting hydraulic lines plagued its operation from the beginning to the end. In its place, the sledgehammer-driven alternative, called plugs-and-feathers, although less sophisticated and requiring considerable physical exertion, proved the more reliable (see Fig. 9). This

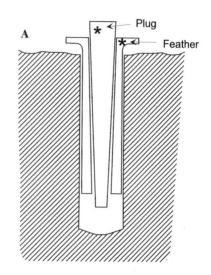

Figure 9. Plugs-and-feathers is the name given to a system of three wedges that are placed into a hole in a rock in order to break it. **A.** When the central wedge, the plug, which is made of hardened steel, is struck by a sledge hammer, the outer two wedges, the feathers, which are made of a softer steel, exert a tremendous force on the side of the hole. **B.** A line of such plugs-and-feathers can neatly break a massive rock so that it can be used as building stone. This simple technique of splitting rock has been used for millennia.

B

often proved to be the case at Dinosaur Cove over the years. What was simple, even if more physical effort was needed, was often more reliable and efficient. But not always. The "not always" is the reason that one can never afford to close one's ears to suggestions for improvement.

Once an advance of the tunnel had been made, it was frequently possible to use rock breakers in conjunction with the rock drills and wedges to widen out the tunnel. This combination of techniques would continue to be used exclusively for another two years.

After a week, it was possible for John McAllister, the mine supervisor, to duck his head and enter the tunnel to announce humorously that at last he was working underground. By then the tunnel had been advanced 1.3 meters (4 feet), but that proved to be enough. When the crew took up the floor where the fossiliferous layer 20 centimeters (8 inches) thick occurred, they found that the highest concentration of fossils was midway between the entrance and the back wall. It was evident that the axis of the ancient stream channel had flowed parallel to the present rock face, so the strategy was to widen the tunnel rather than go deeper.

A few obvious dinosaur limb bones were collected from the tunnel floor. Some displayed the same fourth trochanter that characterized the first recognized ornithischian femur found in 1978 near Eagle's Nest 300 km (180 miles) to the east (Plate 2A). It was gratifying that three years after the area had been officially named Dinosaur Cove, we had finally found undoubted dinosaur bones there.

Although children are fascinated by dinosaurs, many soon tire of watching the process of actually digging them up. This is particularly true when there are no dinosaur skeletons to be seen. Our daughter, Leaellyn, then six years of age, was no exception. After patiently observing what was going on for some time, she expressed a strong desire to go hunting for seashells. So off she and Pat went to do just that. While Leaellyn searched for seashells, Pat scoured the rocks on the west side of Dinosaur Cove. Although others had prospected there before and seen nothing, she found two bones, one of which was rather large.

This second site was quickly designated Dinosaur Cove West, and the original site was renamed Dinosaur Cove East. It was obvious that there were several square meters of rock exposed at Dinosaur Cove West that could be reached without tunneling. Excavating there commenced at once. By the end of the final week, about equal quantities of fossils had been found at the two sites.

In the second and final week, the frequency of mechanical breakdowns and other delays declined and more steady progress was made.

Two days before the planned end of the excavations, the sea had been running high all day. A television camera crew arrived and wanted

Figure 10. Betty Thompson recovering a rock breaker that had been swept away by the high sea.

to photograph operations at night. They were down on site at about 10:00 P.M. when Tom received a telephone call from there saying that the lights had gone out and a hydraulic line of the Darda had burst once again. Although supposedly the sea should have been withdrawing as the low tide approached, water was, in fact, swirling around the crew's ankles. The caller asked what to do, and Tom said in view of the fact that the sea was still unexpectedly up and the gear had failed, they should leave at once. The camera crew provided light for a few more minutes while filming, and then everyone left the site, wading through the surf that should not have been there.

The following morning, Tom was the first person on site. As he approached, a steady whistle greeted his ears. Compressed air was screaming out of a broken hose. The air receiver, which had solved the problem of air pressure drop, was gone. A high sea had swept it off the ledge about 2½ meters (8 feet) above normal high tide level, where it had been placed nine days before. Only the pressure gauge was left; the air receiver was lost to the sea.

Rounding the corner to the entrance to Dinosaur Cove East, Tom saw that the wooden portico, the 250-kgm Darda, and all the other pieces of mining machinery were gone. Thank goodness the electricity had failed the previous evening and the crew had pulled out nine hours before rather than continuing to work all night as planned. Would there have been a warning in a steadily rising sea that would have caused the crew to leave before the waves reached the ferocity capable of doing all that damage? Or did a single high wave cause all this mayhem? If the workers had been there, would they have been able to save themselves? We shall never know.

From that time onward, work at night ceased and has been permitted only once after this. That "once" was an exceptional circumstance years later when the behavior of the sea in Dinosaur Cove was extremely familiar to us. That day there was a hot north wind blowing, the sea was languid, and we knew by then that such conditions would prevail for hours. At such times, one could paddle a rowboat into Dinosaur Cove and safely tie it up to a rock at the edge of the shore platform, whereas at most times to do so would quickly convert the craft into matchwood!

Most of the equipment washed away by the high sea was found in rock pools or buried under tons of debris near the head of Dinosaur Cove. With trepidation, Tom telephoned Atlas Copco, asking what, if anything, could be done to salvage his gear. The Atlas Copco representative told him not to worry because if equipment of that kind could not fall off a five-story building and survive, it would not last in the hire business. "Simply shoot a bit of oil through the airlines and she'll be right, mate," was his advice, and when his directions were followed, "she was right."

In the morning, the sea remained so high that salvage operations were halted for a time. By the afternoon, it was possible to work a bit more at Dinosaur Cove West. The following day was the beginning of the take out, the organization of equipment and sacks of fossiliferous rock so that the helicopter could be loaded quickly on the following morning. On the final morning, the Surf Life Saving Victoria helicopter arrived on schedule, and the gear was brought up to the camp site as the tents and cook shack were being dismantled. By evening, the most frantic sixteen days of our lives were over, the drive home was under way, and plans were already half formulated for a return to Dinosaur Cove.

The Rubicon had been crossed. After recovering several dinosaur bones and a tooth from two sites at Dinosaur Cove in February 1984, we were convinced that deposits existed there that could yield enough fossil bones to make systematic excavation worthwhile. We were also sure that the methodology to work this site was within reach, both technically and financially. It was disappointing that the bird and mammal fossils had failed to turn up, but at least the sizes of some of the fossils indicated that with further work, they might be found. Unbeknown to us at the time, twelve years were to pass before the first one would turn up.

3

Back to Dinosaur Cove

A Concrete-Pouring Bee

How to process the sacks of fossiliferous rock that had been so laboriously collected at Dinosaur Cove was a problem yet to be solved. The rock needed to be laid out to weather in a place where the fragments would be safe. Ideally, this would be on an extensive expanse of concrete or bitumen (asphalt), but none was readily at hand.

Turning once again to the Friends of the Museum of Victoria (their new name was a result of an administrative shuffle), we sought their support to pour concrete pads in our backyard in Emerald. This was a singularly unsatisfactory place to carry out such a plan, because the only area available was rather steep. Fortunately, in imitation of the Incas, a previous owner of our land had terraced part of it in an unsuccessful attempt to grow strawberries. Over a weekend, an enthusiastic crew poured strips of concrete about 1 meter wide on these terraces, using wheelbarrows and small electric cement mixers. Their efforts provided enough area to lay out all the fossiliferous rock collected the first season.

Besides supplying the labor to carry out this project, one of the members of the Friends was instrumental in obtaining the sand, gravel, cement, and wire mesh needed for the work. This was our first introduction to John Herman, an individual who would play a vital logistical role in the years to come.

The Excavation of 1985

Returning for a second season at Dinosaur Cove was a totally different experience. Gone were the uncertainties of approach and whether fossils would be found at all. On the other hand, gone, too, was the enthusiasm of many organizations and people who had provided so much assistance the first time around.

During the second season, the crew would never number more than a dozen, and at one stage, it got down to four. Rather than a mere sixteen days, we spent seven weeks excavating during the second season. Instead of helicopter support, a donation of bamboo poles from the Royal Botanical Gardens, which were carried on the shoulders of as many as six workers, provided the means to transport heavy equipment into and out of Dinosaur Cove.

Figure 11. Moving heavy equipment in a time-honored manner at Dinosaur Cove, 1985.

Initially, all effort was concentrated on Dinosaur Cove West, because the fossiliferous rock there was so easily won. After a week we had collected more specimens there than during the sixteen days of the previous year at Dinosaur Cove East. But we had also excavated all the readily available fossiliferous rock. Faced with removing one or more meters of overburden to get at the additional fossiliferous rock there, Tom chose instead to ask some of the crew to return to Dinosaur Cove East.

Digging at Dinosaur Cove East showed that the fossil layer originally discovered was only the top of a sequence of such layers separated by a series of barren sandstones. As we dug deeper, we found that
the lower fossiliferous layers continued out onto the shore platform, so
that tunneling was not needed to reach them. By the end of the season,
not only had we found several hundred bone fragments (including a
lower jaw of a hypsilophodontid dinosaur) in these newly discovered
lower layers, but we also knew that bones occurred over an area of 35
square meters (42 square yards), about ten times the area uncovered in
the tunnel excavated the previous year.

Furthermore, the bottom of the fossiliferous sequence rested on a
massive claystone. That was just what we had seen at Dinosaur Cove
West. As soon we realized there was a common pattern, we fanned out
over Dinosaur Cove to have a careful look to see if there were other
places that had this same sequence of a sandstone with chunks of clay
in it (the fossiliferous layer) resting on a thick, uniform claystone re-
gardless of whether there were fossils showing on the surface of the
sandstone with clay chunks. Within twenty minutes, Michael Whitelaw,
one of Pat's students at Monash, had not only found such a sequence,
but by digging into the sandstone with clay chunks, had recovered three
fossil bones, although he had noticed none on the surface.

There were now three fossil-bearing sites in Dinosaur Cove. Dinosaur Cove West and Dinosaur Cove East were both west of the new
site. Rather than rename Dinosaur Cove East a second time, the newly
discovered site was designated Slippery Rock, because algae clinging
to the rocks on which we stood to dig for fossils made standing and
walking in the area a precarious exercise—it was indeed a slippery rock.

During the field season, as the fossiliferous rock was excavated from
the various sites, those pieces which lacked any outward sign of bone
were placed in hessian (burlap) bags. Each bag was weighed as the rock
was added until about 20 kgm (44 lbs.) was reached. This was done so
that when a helicopter took them out of Dinosaur Cove it would be
possible to quickly make up a load of a known weight. The reason we
saved the rocks was that they would be spread out to weather for a year
and then broken up further. Because no helicopter support was available, by the end of the field season, 368 bags, or approximately 7,360
kgm (16,200 lbs.) of rock, were stored under an overhanging ledge
awaiting the day when a helicopter could be obtained to transport them
to the paddock near the crew's campsite 2 km (1¼ miles) away.

For four months, no offer of helicopter support was forthcoming.
Finally, Race Mathews, who happened to be both the Minister for the
Arts (which included the Museum of Victoria) and the Minister for
Police, attended a function at the museum. While talking with Tom, it

developed that Race had a soft spot in his heart for dinosaurs. Help was at hand.

Any activity in the Otways is always subject to the vagaries of the weather. Planning to do something there in midwinter on a particular day is definitely tempting the Fates. But on the appointed day in August 1985, the weather was ideal. Seventy high school student volunteers from Apollo Bay formed a human chain to pass the sacks from their repository under the ledge out to a point on the shore platform where the helicopter could land. Accompanying the police helicopter (which Race Mathews had organized) was a state emergency helicopter because it was thought that Dinosaur Cove was too far from Melbourne for the police helicopter to go unescorted. Not only that, but three television stations sent their helicopters as well, so at times it seemed as if Dinosaur Cove was having a scene shot for *Apocalypse Now*. A helicopter would fly in and be loaded with seventeen bags and take off and the next one would soon follow. Two helicopters flying in a circle completed the whole task in half an hour.

The fact that none of the television stations which sent helicopters more than 200 km (120 miles) to cover the lift-out would consider having their own helicopters do the job when we asked for help gave us pause and made us wonder about the way the human mind works. Oftentimes it seems that much more money is spent on covering an event than the event itself costs. There is a message there somewhere about the absurdity of modern society, but we have not figured out exactly what it is.

The Mayor of Dinoville

If this were a novel rather than a chronicle about actual events, we would never have had the imagination to invent John Herman. In 1985, as he observed the Herculean task of moving heavy equipment in and out of Dinosaur Cove using bamboo poles, he remarked time and again that what was needed was a flying fox or aerial tramway. Tom most heartily agreed and said, "Why don't you build it?," while Pat had visions of the magnificent disaster involving a flying fox in the movie *Zorba the Greek*. With most people, that would have been the end of the matter. It is easy to see an obvious need like that, but quite another thing to implement it.

What is a flying fox or aerial tramway, this device that was to so greatly facilitate the operations at Dinosaur Cove? The best way to visualize it is to describe the one John Herman built. At the top of the cliff on the east side of the cove, John built a steel tripod about four meters (13 feet) high. That was the anchor point for a steel support cable that was 1,000 feet long (305 meters) and about 1 cm (²/₅ inch) in diameter. The

Figure 12. A. The basket of the flying fox being loaded with equipment for the trip to the top of the cliff at Dinosaur Cove.

other end of the cable was attached to steel bolts grouted into holes drilled into the shore platform. A wire basket was constructed which was suspended from the steel support cable by chains that attached to a series of pulley wheels. Items to be taken into or out of Dinosaur Cove were either put inside the basket or chained to the outside of the basket. A trap door was fitted into the bottom of the basket. When the load consisted of sacks of rock coming out of the cove, the basket could be quickly unloaded at the top by kicking the lever, which released the trap door (Fig. 12B). The basket was lowered down the steel support cable by reeling out a flexible light steel tow cable that was wound on a drum turned by a pneumatic motor which was in turn powered by an air compressor. Because the bottom of the supporting steel cable was 90 meters (300 feet) lower than the top, all that was needed to make the basket descend was to unwind the tow cable from the drum which was at the top. To pull up the basket was a simple matter of reversing the direction of the drum's rotation. Six minutes were required to haul up a load.

Like any such system, there were teething problems. The most serious was that when the basket neared the shore platform end on its

Figure 12B. A load of sacks of fossiliferous rock being removed from the basket of the flying fox at the top of the cliff by the simple expedient of kicking a lever which released a trap door. The cable was exactly 1,000 feet or 305 meters long as determined by an Australian Army surveying team. The trip to the top took six minutes. Slow though that was, it was a vast improvement over bamboo poles. It could easily carry 200 kgm (440 lbs.). A light tow cable was used to pull the basket upward. Powering it was a compressed air–driven winch motor supplied by Atlas Copco. The driving force for the descent was gravity alone. However, the descent was controlled by the winch motor. At the bottom, the support cable was anchored to the rock of the shore platform by the simple expedient of drilling a few holes and grouting in some rock bolts. At the top, however, a heavy tripod about 4 meters (13 feet) high had to be built and then stabilized by steel cables anchored to the ground to withstand the strong lateral force of the 305-meter-long support cable.

descent, the tow cable went slack and often became tangled in the bushes below. To overcome this difficulty, we used a system of three pulleys spaced at intervals of about 25 meters (80 feet). They had to be lowered after the basket had started its descent by hand-winding a second drum on which there were three additional tow cables.

John is a man of many talents. One of them is that he is mechanically gifted. Another is his acquisitive nature, which has led him to accumulate literally warehouses full of old machinery and this and that, the result of his having attended countless auctions. His endless energy and a generosity toward others (which seems to know no bounds) are a formidable combination; when he took on the task of constructing a flying fox for Dinosaur Cove in time for the 1986 field season, it was done. Not only did he design it with the aid of Bill Loads, who checked the plan, but he managed to obtain donations of those pieces of equipment that he could not supply from his own stockpiles. Finally, with the help of the crew at Dinosaur Cove, he built it, and it worked with relatively few setbacks.

Over the years, when someone would make an offhand remark, such as the time Pat commented about how desirable it would be to have one portable building in which food could be served away from the flies and another building in which to cook, John would show up a few weeks later, unannounced, with just such structures on his truck all ready for assembly in an hour or two. John constructed so much of the equipment around the camp, or at least supplied the building materials, that he became known as the "Mayor of Dinoville," a nickname for the camp. We often wondered whether if John had been given his head there would not only have been paved streets lined with electric lamp posts in the camp but a tram service and a postal code for Dinoville as well!

The Excavation of 1986

Just as a whole book provides much more information than a single page chosen at random, whole skeletons are much easier to identify than isolated bones and teeth. Given a choice, dinosaur workers concentrate on those parts of specimens they know to be from the most complicated regions of the skeleton. This concentration means that skulls tend to be described in detail, whereas ribs rarely are. A rib is basically a rod with only a limited range of variation, while, in sharp contrast, skulls vary much more in form. There is more biological specificity in the more complex parts. Or, to put it in terms of information theory, there is a lot more information in a complicated skull than in a simple rib.

Because of the fact that the majority of dinosaur remains from Victoria are single bones and teeth, their identification has proven to be extremely difficult. However, the very incompleteness that can be so frustrating does serve an important function: it prevents you from doing the same kind of study over and over again. In order to get the most out of the specimens being analyzed, you must constantly be thinking of new approaches to gain the most insight about them. The approaches are partially dictated by which parts of the bone happen to be preserved and what postmortem damage they have undergone. If we had used only the features that are common to all specimens in carrying out our studies, we would have had very little to analyze. In a field where missing data is the rule rather than the exception, we had to take advantage of unique circumstances of fossil preservation we encountered. To do that, we had to be ever on the alert to the possibility of such occurrences. As a consequence, we sometimes gained truly unique insights. Those are among the sweetest fruits of the scientific enterprise.

When one finds primarily isolated limb bones or fragments of skull, as we did, it is almost impossible to identify most specimens by referring to published descriptions. The quite understandable literature bias toward giving detailed descriptions of those parts of skeletal anatomies which are most informative worked against us. Although the necessary information for their recognition often exists in individual bones, they are seldom described and illustrated in enough detail to make it possible to identify them solely by reference to scientific publications. Because the published information about dinosaurs could not help us identify these fragmentary specimens at a satisfactory level of certainty, we found it absolutely essential to go where the most extensive collections of dinosaurs are located and examine actual specimens in order to discover what the features are of a particular fossil. Recognizing features in the bones we happened to be studying that often had never been considered of significance before sometimes enabled us to unexpectedly identify their affinities.

Because Australia has very few dinosaurs in its museums, we have had to examine collections in museums that are overseas. In late 1985 we had our first opportunity to do this since we began work on Victorian dinosaurs. Not only were we able to improve the identification of many of our fossils, but we also had a chance to speak with a number of people with similar interests.

In particular, at a meeting where Tom gave a summary of the polar dinosaurs from Victoria, the preceding paper was about the polar dinosaurs from Alaska that had been recovered during the previous northern summer. It seemed at that time that discoveries about polar dinosaurs were undergoing an information explosion in both hemispheres. The possibilities of comparing the evolution of dinosaurs in the two polar regions would eventually lead to exchanges of scientific workers between the two hemispheres.

Of necessity, this overseas visit overlapped with the beginning of the 1986 season at Dinosaur Cove. In order for the work to go ahead, the organizing and initiation of that field season was left in the capable hands of Michael Whitelaw, who had discovered Slippery Rock the previous season, and Elizabeth Thompson, then Tom's assistant at the Museum of Victoria. Once the fieldwork was under way, Michael was ably assisted by Keryn Walshe, who had participated in the previous excavations and was extremely proficient in the use of even the heaviest air-driven tools. Subsequently, she worked with us in Central Australia. Even after we returned in late January 1986, she and Mike took care of most of the day-to-day operations.

Figure 13. Volunteers were found for the 1986 excavation with a placard reminiscent of an army recruiting poster.

Many people like dinosaurs and digging for fossils. It is fortunate that they do, because without literally hundreds of such people who have volunteered to help over the years, the polar dinosaurs of Victoria would no doubt still be in the ground. People who love both dinosaurs and digging often offer their services as volunteers at a dinosaur dig for a combination of reasons. Many have had a lifelong fascination with fossils and fossil collecting. Added to that is the attraction of meeting and working with people who are similarly inclined on a serious scientific project. Finally, for non-Australians, there is the attraction of visiting Australia—the Antipodes. Once motivated, some kept coming back year after year.

Unlike people who are paid, there is no monetary string holding a volunteer to the project. If they are treated in a way they see as unfair, they're gone. We were able to keep enthusiasm high by allowing individuals as much freedom as possible in choosing their tasks (at the same time being sure that all aspects of the job were done by someone) and by providing the necessities of life. Another important aspect of keeping up the fervor of volunteers was making sure that each person understood both where their contribution fit into the overall task and, for those interested, what the developing scientific picture was that their efforts were contributing to. Many discussions were held in the evenings on these topics in the big communal tent after tea (as the evening meal is called in Australia). These discussions not only informed the volunteers but served to sharpen our own thinking as well.

After 1984, when there was always a cadre of old hands who knew the ropes, the first morning of a field season was always a very casual affair. Everyone would go down to the site and because the old hands knew what had to be done, they self-selected for the various jobs. The new chums/recruits often were a bit bewildered by the apparent lack of structure to the group. But they soon asked a question or two of the old hands and found a job that needed doing. They then either self-selected for that job or asked around and found another one they liked. In this way people found a task to their liking with a minimum of fuss. Many of the best volunteers had a feeling of general responsibility for the project and would take on all but the most onerous task if they saw a need for something to be done. Only rarely did Tom have to direct someone to do a particular job. When it was possible, he would work by himself while keeping an eye out that what needed to be done was in fact being done. Because child care and camp cooking were still major responsibilities, Pat usually spent most of the day in camp, especially in the 1986 and 1987 seasons.

Very few people would take on one most onerous task unasked, and even if asked, people would universally refuse. This was to return to camp and get a piece of equipment that had either been forgotten or was needed on site because of a breakdown. If the item was not obtained, the result would typically have been that work would not go forward that day or would progress rather inefficiently. There was a good reason for their refusal: the return to camp entailed climbing a steep slope 90 meters (300 feet high). Tom generally wound up with that job. It was a good day when he had to make only one climb out for a piece of gear.

Volunteers for the work at Dinosaur Cove have largely been drawn from four sources. Some people contacted us after learning of the work through the Friends of the Museum of Victoria, others learned by word of mouth, still others learned by way of the media, and key supporters actively recruited university students. A fifth group was added in 1986.

Earthwatch Volunteers

Earthwatch is an organization based in Massachusetts, U.S.A., which, in a role somewhat akin to a matchmaker, gets people who want to participate in scientific fieldwork together with scientists who need their help. The money that the volunteers donate to Earthwatch is divided between them and the scientist's project. The balance retained by Earthwatch is what makes it possible for that organization to function. The Earthwatch volunteers got themselves to the Melbourne airport and assisted the project by providing not only their physical effort but cash as well. Not knowing what such volunteers would be like, we recruited only a small group to come the first season.

Figure 14. The inaugural Earthwatch crew, 1986.

On average older than the other volunteers, Earthwatch volunteers proved to be excellent help. We soon learned that people who pay to work are determined to earn their calluses. Whereas about 70 percent of other volunteers made a substantial physical contribution to the work, among Earthwatchers 95 percent did.

As a catalyst for the remainder of the crew, their positive effect was quite profound. A hulking university student who was feeling tired toward the end of the day as he lugged sacks of rocks would invariably find hidden sources of energy he did not know he possessed when a female Earthwatch volunteer old enough to be his mother carried an equally heavy load.

Unfortunately, this project never attracted enough Earthwatch volunteers, so the organization never reached the break-even point in its support. The cost to Earthwatch of publicizing our dig and recruiting volunteers for it was greater than the amount of income they received from the volunteers. Therefore, after 1991, they terminated their participation. By that time we were so impressed with the inspiring contribution of Earthwatch volunteers that we offered in the future to take those they could recruit and let Earthwatch keep all of the fee the volunteer paid. However, their charter did not permit that unusual arrangement, so our association had to cease, but we hope not forever. The end of the relationship with Earthwatch did not end the relationships with several of their volunteers, who have since returned to Dinosaur Cove on their own or have helped in other ways.

Two of the first group of Earthwatch volunteers were Herm Seibert and his son Steve. Herm is a retired engineer and Steve is a lawyer who currently serves in the cabinet of the governor of Florida. Herm has been on many other Earthwatch projects and is particularly fond of an archeological site in Thailand, to which he has returned a number of times. Both Herm and Steve returned together five years after their first visit and helped start the construction of "The Great Wall of Dinosaur Cove." Herm returned a third time when the primary operation had shifted from Dinosaur Cove to Flat Rocks 300 km (185 miles) to the east.

Among the other Earthwatch volunteers was Bob Hodge, a retired junior college science teacher from Fredericksburg, Virginia, who has now moved to Emporia, Kansas. Many people like to break up rocks and look for fossils, but none can possibly exceed Bob at that skill. On days off, one would often find Bob off cracking rocks by himself and whistling to himself as he did so. Nothing made him happier than to find a bit of fossil bone after hours of seeing nothing but barren rock. He came back several times and on the last trip was accompanied by his wife, Lois.

Figure 15. Herm Seibert.
Photographer: Peter Menzel.

There was also Bill Hopkins from Anchorage, Alaska; before retirement he was executive director of the Alaska Oil and Gas Association. Not only has he assisted the polar dinosaur project in the southern hemisphere, but he has also been instrumental in helping us try to get a project off the ground to excavate polar dinosaurs on the Colville River of Alaska (see Chapter 12). Allan Fraser, a physicist who formerly worked for the U.S. Navy developing electronic black boxes for use in submarine warfare, has returned to help more times than any other overseas volunteer. He once joined Tom in Patagonia to search for Early Cretaceous dinosaurs there (see Chapter 12). He is also using his engineering knowledge to help us understand the structure of the skull of the giant extinct marsupial *Diprotodon* (see Chapter 7). John Wilson is an expert mechanic who originally came from Oregon. When in Australia for the 1986 dig, he met his future wife, Helen, a student at Monash University and member of the field crew. They both returned for the 1989 dig. John was a master rock driller, while Helen was an expert at cracking rocks and finding fossils. Now married, they have three children and live in Mount Martha, a town about 70 km (40 miles) southeast of Melbourne.

While an interest in dinosaurs and fossil-digging initially drew people to offer their services as volunteers, many of those who returned again and again found the social life of the dig appealing. There was a camaraderie among the crew, and many lasting friendships were formed that have persisted for a decade or more. Five couples that we know about married after meeting at and participating in the digs. The frequency of the formation of such liaisons led to wry speculation that perhaps dinosaur-digging was merely a cover for a clandestine matrimonial bureau.

After the discovery in 1985 that there was reasonably accessible fossiliferous rock under the shore platform at Dinosaur Cove East, half of the effort during the eight-week 1986 excavation was devoted to recovering specimens from there. Numerous limb and pelvic or shoulder girdle elements turned up, as well as a few teeth.

In 1986, for the first time, a major effort was made at the Slippery Rock site, discovered in 1985. Because there the fossiliferous layer occurred on a vertical face about half a meter (20 inches) above the shore platform and the underlying claystone was somewhat easier to dig out than sandstone, undermining was the first approach we took to excavate fossils from there. Suspending a rock breaker by a rope to make it possible to operate it horizontally, claystone was removed from beneath the fossiliferous layer over a distance varying from 2 to 5 meters (16 feet) along the face and up to half a meter (1½ feet) underground. Then a series of holes was drilled just above the fossiliferous layer at intervals as close as 20 centimeters (8 inches), into which as many as twenty sets of plugs-and-feathers were placed. In this way, large slabs of the fossiliferous layer were broken free by being dropped into the space where the claystone had been. These were then broken up and searched for fossil bones.

On the first day at Slippery Rock the crew was almost ecstatic because as soon as the initial block was broken free, six bones were visible, and by the end of that day fifty-one bones which were good enough to keep had been found.

By the end of the season, more than three times as many bones had been found as the previous year, about two-thirds of which came from the Slippery Rock site. While it was clearly desirable to continue at that spot, a change of approach was just as clearly needed—the undermining of the cliff over a distance of 9 meters (30 feet) could not continue inward indefinitely. The narrow slots dug in the claystone when the advance was limited to 50 centimeters (1 foot 8 inches) would clearly have to be replaced by spaces high enough for an adult to stand in. If they had been dug in the claystone below the fossiliferous layer, the spaces would have been below sea level much of the time.

We decided to dig tunnels above the fossiliferous layer and then work down to it, as we did initially at Dinosaur Cove East. The effort of doing this by mechanical means alone, however, was just too daunting. We resolved to investigate the use of explosives for this purpose, something we had firmly resisted up to this time. Previously, we felt that there was so much else to learn about this kind of work that to add the complications of using explosives was trying to do too much at once. But now we had a group of people experienced in the other aspects of working at Dinosaur Cove, and we had acquired the equipment to do so. Tom concluded that 1987 would be the right year to learn how to blow up rocks.

In deciding to take this course, Tom deliberately made a second decision. He, personally, would never learn anything about the actual wiring up and firing of explosives. Just as it is best to either fly frequently (in order to be an excellent pilot) or not to fly an airplane at all, it seemed to him that he would never do enough underground shot-firing to gain the level of skill needed to be able to do it safely on his own. If there was always going to have to be someone qualified around to supervise him, then why learn to use explosives at all and ever be tempted to try when a qualified individual did not happen to be around and blasting needed to be done?

Boom! The Excavation of 1987

Dinosaur Cove and all the other places where dinosaurs have been found in Victoria were located in areas of coastal shore platforms under the control of the Department of Conservation and Natural Resources. The Department required us to get permits each year to carry out excavations. We not only had to seek their general approval, but we also had to reach an agreement with them about the methods we would use.

These requirements ensured that this precious coastal scenery was not unduly harmed and also provided a safeguard for the scientific value of the sites. The rangers in each area were keenly aware of where the fossil sites were. This meant that large-scale, unauthorized private collecting of these localities, which would deprive the general public of what was rightly theirs, was not likely to continue for long without rangers becoming aware of the situation.

After a number of years, we had come to an agreement with the Department of Conservation and Natural Resources about an informal schedule for the steps in the permit application procedures. The permits for 1987 were different for two reasons. First, we had to consider the effects of the use of explosives. Second, for unknown reasons, Tom was close to being blind for a month and a half in late winter and early spring, throwing the schedule into serious disarray. This situation nearly caused the cancellation of the 1987 excavation. However, we felt that if the project lost momentum, it might never start again.

Normally, the permit for an excavation that began just after the new year would have been submitted in mid-October. However, in 1986, our permit application did not go in until mid-December, little more than a fortnight before it was needed. Arnis Heislers, the official then in charge, promised to do all he could but would guarantee nothing; the Christmas season had begun, and the minister and senior officials who would approve our application were about to go on leave. Because of Arnis's untiring help, the day Tom departed to start the 1987 field season, the permit was in his hands.

In many other (less extreme) ways, the Department of Conservation and Natural Resources consistently assisted this work, which, without their enthusiastic support, could not have gone ahead. Their continued surveillance of the sites has also been a long-term blessing.

In those last weeks before Christmas of 1986, many other critical problems had to be solved. For more than six months, Tom had what seemed like endless conversations with people in the mining and tunneling industries, seeking one or two individuals who were qualified to act as a mine manager and underground shot-firer and were willing to work for free. The normal cost of hiring such people was so far beyond our means that it was pointless to try to attract anyone but a volunteer. In desperation, in mid-December he took out an advertisement for such volunteers in *The Age* newspaper, which was paid for by our long-term supporter, Bill Loads.

Figure 16. Advertisement from *The Age* newspaper, December 1986.

Several individuals came forward. Eric Leach was chosen to be the mine manager and another individual was selected to be the underground shot-firer. Several other people who had come forward expressed a willingness to join the excavation as ordinary volunteers, and they were welcomed, for the more people with tunneling or mining experience, the better it would be, given the new areas of expertise that were going to be needed in this operation. After we had made these selections, we had overcome the major obstacles to the onset of our operations for 1987.

Just to be as sure as possible that the initiation of blasting at Dinosaur Cove would not have an adverse effect on the immediate environment, Tom asked two colleagues at the Museum of Victoria, Dr. Gary Poore, an expert on marine crustacea, and his assistant, Helen Lew Ton, to visit and assess what the impact was likely to be on the marine life there. They did so and pointed out that the environment was subject to so much natural turbulence that blasting on the scale planned would cause no measurable effect.

Once the field party arrived on site, we needed eleven days before drill steels could touch the rock to begin the tunneling. We had to erect the temporary buildings in the camp; install the flying fox; put the compressors, air hoses, air receiver, and pneumatic tools in position; bar down loose rocks on the Slippery Rock site; and install the telephones. We also had to dig the powder magazines well below ground and establish a high-pressure water system. The high-pressure water system was necessary to minimize the dust from the underground drilling. With it in place, a steady stream of water flowed through the drill steels when they were in operation, turning the rock powder to slippery, oily mud, which, although unpleasant in its own right, was highly preferable to choking dust.

Two powder magazines were needed, one for the detonators and the second for the explosives, each in a hole of its own. The site we chose for the magazines had soil the consistency of excellent adobe. In subsequent years, this meant that even in the rainy climate of the Otways, there was little collapse of the walls and, therefore, that little maintenance was needed. In a burst of enthusiasm by one crew member, the first hole was arduously dug almost entirely by hand by a progressively more disgruntled group of volunteers until David Denney, the farmer on whose land the camp and powder magazine were located, quietly asked if his backhoe might be of some use.

From the beginning of the tunneling work at Dinosaur Cove until it ended, David and his wife Winsome helped the project in a myriad of ways. They not only allowed us to set up camp on their land year after year but were often the first people we turned to when a difficult problem arose. David was an endless source of ideas about how to solve the many technical difficulties that arose with regard to all the machinery we used to excavate the fossils at Dinosaur Cove. In addition, he had an incredible accumulation of nuts, bolts, old tools, and machinery, out of which many pieces of gear were designed and built, often late at night, to overcome some problem which was holding up progress in the excavation.

Figure 17. Dave Denney (left) and Tom. *Photographer: Peter Menzel.*

Like many of the other people who lived in the Otway Range near Dinosaur Cove, they were not only very interested in the project but integrated what they learned about it from us with their detailed knowledge of the general natural history of the area. In our long discussions during the evenings with them, we gained much local knowledge that was relevant to what we were trying to do.

Because none of the people involved had ever excavated holes for a powder magazine before, we followed the regulations to do so to the letter and interpreted them conservatively. As a result, when Patrick O'Neill, the Victorian Government Inspector of Mines and Machinery and Inspector of Occupational Health and Safety came to check it, he merely shook his head and muttered something to the effect that "that will *certainly* do."

Pat O'Neill had a long association with the project in his capacity as a Department of Labor inspector. He not only made sure that safety standards were met but was ready, at the drop of a hat, to assist whenever his help was needed. On one occasion, after a professional hard-rock miner casually visiting the site told us that a particular situation was quite dangerous, Tom raced to Lavers Hill, the nearest town, and telephoned Pat. It was a Saturday, but O'Neill was at Dinosaur Cove the following day. He quickly reassured Tom that nothing was wrong. Had the tunnel been 500 meters underground it would have been another matter, but a different set of rules applies to shallow tunnels than to deep shafts. He pointed out that just as there are things that can be done at shallow depths which are foolish at deep ones, there are things possible at deeper depths which are not safe in shallow tunnels. Experts in one kind of underground work are not necessarily experts in the other.

Without ever visiting the site, other government officials were dismayed at the stories they heard about what was going on at Dinosaur Cove. Rather than taking the trouble to personally visit the site, some tried to shut down the operation. Pat O'Neill was able to deflect these

Figure 18. Pat O'Neill (right) and Tom in Dinosaur Cove. (Courtesy of *Time* magazine)

Figure 19. The Lavers Hill store. Thirteen kilometers (8 miles) from camp, this was our closest source of supplies and access to communications. Several thousand dollars' worth of coins were poured into the telephone box at the right over the decade that the dig at Dinosaur Cove was under way. *Photographer: Peter Menzel.*

Figure 20. Joy Wilkinson serving a customer at the Lavers Hill store. Often Joy and her husband Chris helped us out in ways far beyond what one would normally expect of a storekeeper. Without that generously given assistance, many tasks would not have been accomplished in twice the time. *Photographer: Peter Menzel.*

uninformed moves. Tom never learned that this was going on behind
the scenes until years later. Had Pat O'Neill not supported the project
in this way and kept a close personal eye on the work, the work at Dinosaur Cove might well have never reached fruition, and the Flat Rocks
site at Inverloch would have remained undiscovered.

Perhaps no professional tunneler appreciated the dedication and potential value of the volunteers more than Pat O'Neill. Rather than de-
spising them or merely tolerating them, as some did, he was continually amazed at what they would accomplish with the simplest of tools.
He often remarked that you couldn't possibly pay people enough to do
many of the tasks these volunteers were doing for the love of it. They
seemed to swarm chaotically, but with obvious determination, around a
job and eventually accomplished tasks that were often daunting. This
earned the volunteers the sobriquet "The Angry Ants" in Pat's vocabu-
lary.

As the installation process continued, it was evident that there was a
feeling of unease between the new people and the old hands. The new
people with the mining expertise carried with them a professional outlook which did not include women working alongside them as equals.
When women who were old hands were told they should not be operating a rock drill because they were women, their discontent openly
flared. To make a point that women as well as men on this excavation
would do those jobs they were physically capable of, Tom decided to
move in the air receiver, which was needed on site before the flying fox
could be available. Weighing between 200 and 300 kgm (440 lbs. to 660
lbs.), it was a brute to move, not only because of its weight but also
because of its awkward shape. It had to be transported 600 meters (660
yards) from the closest parking area on the edge of Dinosaur Cove, first
250 meters (275 yards) down a steep, slippery track that descended 90
meters (100 yards) vertically and then to its final destination in the cove
itself.

Using bamboo poles, eight of us—five women and three men—hoisted it on our shoulders and started down the steep track. In fits and
starts, we got it nearly to the bottom of the descent. Before us was the
last 20 meters (22 yards), the steepest section of all. Carrying it over
that stretch on our shoulders was impossible because we could not have
maintained our footing, and sliding it over the rocks was difficult because of the roughness of the surface, which caught the air receiver every time it moved a few centimeters. The only solution was to position
one of us below the air receiver, pulling it around or over obstacles while
the other seven restrained it from sliding out of control to the bottom.
In this way it finally reached the nearly horizontal shore platform. The
eight of us then hoisted it again on our shoulders and carried it the rest
of the way. As we approached the new people already working in Dinosaur Cove, we were met by silent stares. This successful maneuver

did much to silence the overt disparaging remarks about what women could and could not do.

It was well that we delayed the onset of blasting for three field seasons. Had all the other technical aspects of the dig been as troublesome as the initial blasting turned out to be, it is questionable whether we would have ever accomplished anything. Initially, we planned to start two parallel tunnels 6 meters (20 feet) apart. We soon abandoned that plan because of the difficulties we encountered. The rock was a heterogeneous mixture of alternating bands of hard sandstones and softer claystones. Added to this, as the tunnels got deeper, the rock was less weathered and changed character in the way it reacted to explosives. Experts from ICI (now Orica), such as the former head of the explosives division, Alastair Blaikie, and the then head of that division, Spence Herd, had rarely encountered such difficult rock to work with in their professional careers.

The first underground shot-firer frequently used as many as eighty sticks of gelignite[17] in a single blast as he tried to get the tunnels started. In doing so, he simply blew the whole face away, shattering the surrounding rock in the process, an outcome that was to plague work there for years to come. Eventually, he was asked to leave, and Eric Leach took on the twin tasks of mine manager and underground shot-firer.

When someone had to be asked to leave, it was important to try and do so in as fair a way as possible. None of those who were asked to go had deliberately done something detrimental. Therefore, we took care to ensure that they understood why their assistance was no longer required. In most cases they understood, and we parted ways perhaps not as close friends but at least as two parties who each understood where the other was coming from. Besides considerations for the feelings of the individual asked to go, there was another, equally important, reason to do this. The rest of the volunteers had to feel that the decision was a just one or their morale would be devastated.

The difficulty with the first shot-firer epitomized the problem that anyone has who is running a project without all the technical qualifications of the people who are working on the project. Without fully understanding all the details of what they are doing, the project manager must be able to identify those individuals who are best qualified. Up until it was necessary to blast, this had never been a problem on the excavation, because as we went along Tom acquired the new skills involved in such things as drilling rock with pneumatic tools. But in the case of blasting, he had to make his judgments by watching performance and assessing it as best he could and by asking technical questions and gauging the response. The art of leadership is really put to the test in cases like this, because the manager must maintain control over the entire project and yet be able to accept and respond sensibly to advice others give that he or she is not technically qualified to independently verify. As with many other undertakings, the seemingly contradictory traits of strength and humility must be combined and made to work well together if a project such as this one is to be successfully carried out.

Except for the first shot-firer, all others were as cautious and conscientious as one could possibly be. However, when one is using explosives, no matter how careful one is, unexpected things will happen that are both thought provoking and unnerving, to say the least. In one instance, four and a half minutes after one shot had been fired, a 100-ton block of rock sheared off the cliff face above one of the entrance tunnels and crashed to the shore platform below. No one was in any danger of being hurt by this fall. This was because it was a firm, official rule of the Victorian government that one waits five minutes after a blast before approaching a site where an explosion has just taken place. But had the fall occurred a minute later, who is to say whether someone might have been hurt or killed?

Another time, a shot-firer carefully surveyed the area to make sure no one was standing where debris from the blast might strike them, blew his warning whistle, waved a red flag as required, and set off a charge. At that stage, the tunnel was about 3 meters deep. This was sufficient to somewhat direct the debris out to sea like grapeshot out of a cannon. A rock from the explosion skipped over the water, crossing the bow of a lobster boat by about 70 meters (230 feet). What the shot-firer had not done was to look over his shoulder out to sea. Had he done so, he would have seen the boat in the process of recovering lobster pots. Accidents are often the result of incredible coincidences; this time no accident occurred only because the boat in question, fortunately, was not where the rock went. It would have been much closer a minute later. After that incident, lobster boats coming into Dinosaur Cove were quite aware of the meaning of red flags and whistles and cleared off hastily when a blast was imminent. We also kept an eye out to sea at such times.

It is a testimony to how small the population of Australia is that this incident had the following sequel. Years later, one of the field crew, Nick van Klaveren, was recounting this story in a pub more than 2,500 kilometers (1,600 miles) away in the gold fields of Western Australia. There was a moment of silence in the crowd of listeners and then one of them calmly and quietly announced that he had been on the lobster boat and had quite vivid memories of the event from a somewhat different perspective!

Not all the blasts were as unnerving as these ones. On one occasion, a wheelbarrow was left in the tunnels between the blast site and the entrance. When the charge went off, it flew out of the tunnel, described

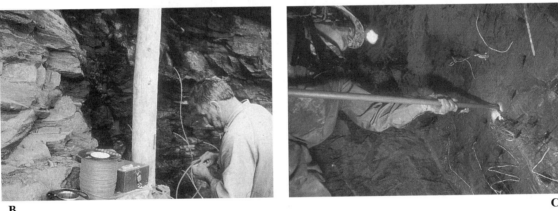

Figure 21. Tunneling with explosives was a laborious three-step process. A. First, a pattern of as many as thirty-six holes was drilled in the rock face using a Panther pneumatic rock drill. *Photographer: Ray Faggotter.* B–C. Second, the explosives were placed in the holes and the detonators for each hole wired together. *Photographer (C): Peter Menzel.* The person wiring the detonators was able to control the blast because each type of detonator had a known built-in delay, a matter of milliseconds, from the time the electrical impulse was sent down the wire from the "Beethoven." Thus, a specific sequence to the explosion was possible. The device which produced the electrical impulse was given the name "Beethoven" because it made such wonderful music, at least to the ears of the explosives experts! D. The shot-firer is holding the Beethoven on his lap in this picture. E. Third, once the face had been blasted, the rock rubble produced was removed or "mucked out," and the cycle repeated. *Photographer: Ray Faggotter.* Here the mucking out is being done using a bogger, or scoop, that is dragged back and forth by an ancient, cranky pneumatic winch being operated by the man standing on the platform. More often than not, it was done with the far more reliable, although more labor-intensive, shovels and wheelbarrows. One meter (1 yard) was a good advance with one cycle. Many times we did far less well than that.

D

E

a graceful arc through the air, turning a slow, partial somersault in the process, and finally landed in the cove, where it promptly sank, fortunately in shallow water. Once recovered, it functioned perfectly well despite a few more dents added to the already numerous ones that previously adorned it.

➥ Eric proceeded more conservatively than the first shot-firer, and was finally able to define the tunnel entrances. The work went ahead. Even with this more cautious approach, problems continued to occur with the blasting. In close consultation with Alastair Blaikie, Eric kept modifying his blasting technique right up to the end of the work in order to achieve the best results because the rock was weathered less and less as we penetrated deeper into the cliff.

Added to these difficulties was adverse weather. Although a watch was maintained for them, high waves on three occasions swamped pumps driven by petrol (gasoline) or electricity that were used to supply water at high pressure to the rock drills, shorting out their circuits. Nearly everyone who worked at Dinosaur Cove for any length of time was drenched at least once, if not knocked down, by a rogue wave.

In later years, the unsatisfactory performance of the pumps we used on the 1987 dig to supply high-pressure water to the rock drills caused us to search for alternatives. The first solution was to put a tank about 15 meters (50 feet) above the level where the rock drills were being used and pump seawater into those tanks from time to time. From there, a gravity feed supplied the drill steels with water. Although 15 meters seemed sufficient to ensure enough pressure for the job, it still proved insufficient. Finally, we built a reservoir at the top of the cliff. With 90 meters (300 feet) of head, there was certainly sufficient pressure. We recharged the reservoir by bringing fresh water in from Lavers Hill. This meant that there was no problem with corrosion of the rock drills and drill steels because of salt, and we could always drink the water as well as use it to make cement.

Many days it was simply impossible to work in Dinosaur Cove because of the high seas. In one instance, four days in a row went by when no one went into the cove for this reason. After the prolonged periods of high seas, it was common to find that car-sized rocks were not where they had been. The crew became so accustomed to high seas that sometimes it was almost too late when finally it became evident to one and all that the time had come to pull out.

In the midst of all this, Eric Leach had the misfortune to slip in the shower, of all places, hurting himself so badly that he had to go home for a number of days. Fortunately, another member of the crew, Don Manning, was qualified to step in as mine manager and underground shot-firer.

We needed nearly a fortnight to complete the first tunnel (or the West Tunnel, as it came to be called), which was located near the western extreme of the Slippery Rock site fossiliferous outcrop. When we dug up the floor of that tunnel where the fossil-bearing rock was expected, we found that only a tiny patch on the eastern side near the tunnel's entrance yielded any fossils. This was quite in contrast to what had been found nearby outside the tunnel the year before. The second tunnel, or East Tunnel, which was located near the midpoint of the fossiliferous outcrop and was completed a fortnight later, yielded somewhat more fossils but only in the 2 meters (2 yards) closest to the entrance.

Tom reasoned that because the bulk of the fossils found outside at Slippery Rock had been located in the area between the two tunnels, that was the most promising place to look further underground. It would not have been prudent to simply strip away all the rock between the two tunnels above the fossiliferous layer, for that would have created an unsupported roof nearly nine meters (30 feet) wide in the weathered rock close to the original surface. Such a structure would likely have been unstable. Rather, Tom decided to cut a third tunnel, the First Cross Tunnel, between the first two. By using this approach, a column of rock was left in place that supported the roof. It was immediately christened "The Pillar." (In 1990, the Second Cross Tunnel would be cut parallel to the first one a further eight meters from the entrances.)

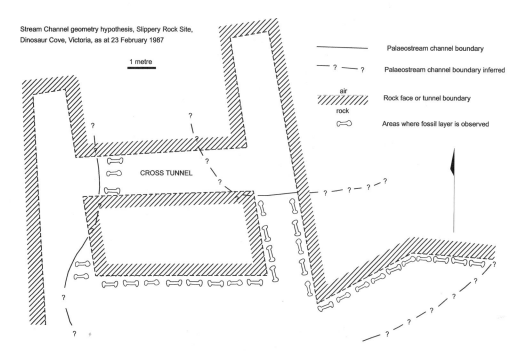

Figure 22. Map of the tunnels at the Slippery Rock site at the end of the 1987 excavations.

Once the First Cross Tunnel was finished, we began to take up the floor where the fossils should have been, if they were there at all, on the forty-eighth day of the dig. The forty-ninth day and the morning of the fiftieth day were equally disappointing; that is when thoughts about the merits of becoming a butterfly collector came to mind. On the afternoon of the fiftieth day, as we dug deeper into the floor at the west end of the First Cross Tunnel, the fossils started to pour out.

Because Victorian public service regulations would not allow anyone but a public servant to drive a state vehicle, the next day Tom had to drive to Colac in the museum utility/pickup truck in order to purchase fence posts to put around the excavation for the powder magazine. For that reason, he was not there when the most exquisitely preserved dinosaur skull was found. Tiny, it was only 51 mm (2 inches) long. The endocast of the brain was beautifully preserved, showing minute structures such as the base of the pineal organ (the pituitary gland in humans) and the cerebral and optic lobes. The bone of this little skull was jet black and contrasted subtly with the dark gray rock in which it was embedded. Later examination of the tiny teeth would show that it was not a bird, pterosaur, small theropod, or a monotreme, as Tom's first thoughts imagined it might be. The size of the brain and delicate sculpture of the skull had initially suggested those possibilities.

His immediate reaction was one of profound indifference. For too long, the work had been so frustrating that of necessity Tom had adopted an attitude of as complete detachment as possible in order to continue what in his rational mind appeared to be the best course of action—but the likelihood of a favorable outcome appeared low. Although he was not then aware of it, he was suffering from clinical depression as a result of his temporary loss of eyesight months before. This delayed response to a traumatic event happens all too often. The following day, more out of a sense of professional duty than with any feeling of accomplishment or joy, Tom deliberately forced himself to take great pains to crate up the rock with the skull in it and send it off to Monash so that Pat could examine it and Lesley Kool could start to prepare it. Years later Tom felt cheated in not having felt a sense of the elation that should have occurred when he first laid eyes on the skull—discoveries like this seldom happen in a lifetime. But without the stoic attitude that numbed him emotionally at the time and which enabled him to continue when the work appeared futile, would the discovery have ever been made at all?

Meanwhile, in the First Cross Tunnel, a string of articulated vertebrae and associated limb bones were found in a series of blocks close to where the skull had been collected the day before. In the darkness and muddy conditions, about half of them had already been thrown outside on the mullock heap (rock pile), from which they were re-

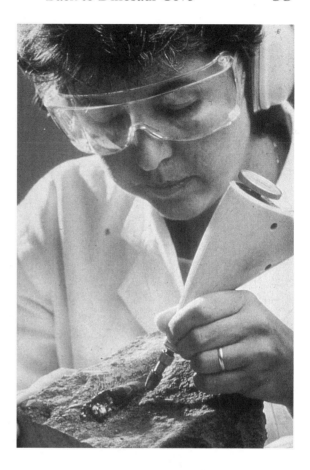

Figure 23. Lesley Kool was a volunteer on the first dig at Dinosaur Cove in 1984. Afterward, she asked Tom if she could become involved in preparation of the fossils as a volunteer. She came into the museum, and he taught her the rudiments in about five minutes. From that beginning, Lesley took the full initiative. She found books about fossil preparation and began to correspond with preparators at other museums, particularly the Royal Tyrrell Museum of Palaeontology in Alberta, Canada. A few years later, she spent a month there learning more about preparation by working on their fossils as a volunteer. By doing that and gaining further practical experience by patiently bringing to light in the Monash laboratory the fossils we were collecting, she fine-tuned her expertise over the years. In this way, she became the key person in our project. Without her, the fossils might never have been freed from the rock.

trieved by the frantic crew members once they realized what had happened. Although it will never be possible to prove beyond all possible doubt that these postcranial elements are part of the same individual as the skull, it is highly likely. The structure of the femur is clearly that of a hypsilophodontid dinosaur, as are the teeth of the skull. Because two or more bones of one individual almost never occur together at Dinosaur Cove, it is unlikely that two individuals both represented by partial skeletons rather than single bones would be found so close together there. Finally, there are no duplicate elements, such as two left femora, to indicate that more than one individual was preserved in that small area.

Originally Tom had intended to end the dig in late February. But with the floor of the First Cross Tunnel paying off so handsomely, to have left the site before all the exposed fossiliferous rock had been taken up would have risked its destruction by erosion before another excavation in two years' time. Fortunately, Atlas Copco agreed to allow their air-powered tools to remain until mid-March if necessary, so the work could proceed in an orderly fashion. Finally the work was finished about ten days later than originally planned, with specimens in hand that shelved all plans to chase butterflies.

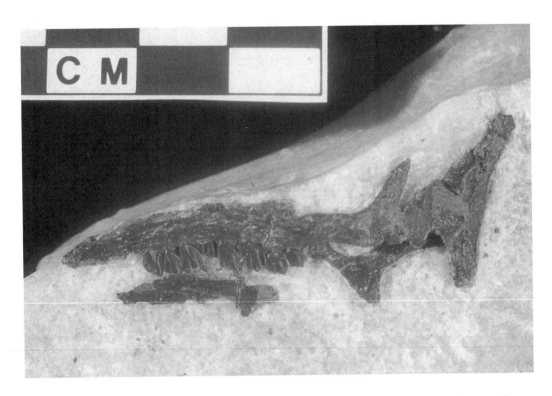

Figure 24A. In order for the outside surface of the skull fragment to be seen, the exposed inner surface that is pictured in Plate 6 was embedded in wax and then the surrounding rock was removed. Because the skull fragment is so fragile, only one side of this skull can be exposed at a time. To view the inside of the skull again would require that the specimen be embedded in wax on the outside and then the inner side exposed by removing the wax that now covers it.

Figure 24B–C. Top of skull of the same individual in Plate 6 and Figure 24A. In the same way as for Fig. 4, this pair of images can be seen in three dimensions. Toward the top of the page is the snout and at the rear a natural mold of the top of the brain is exposed. The left optic lobe is visible in the lower left-hand corner of the brain. A solid line on the left-hand image identifies it. Its large size relative to the optic lobes of hypsilophodontid dinosaurs from lower paleolatitudes suggests that *Leaellynasaura amicagraphica* had unusually high visual acuity.

4

Interlude

By 1986, when it had become apparent that Dinosaur Cove would be a major part of our lives for a long time to come, we resolved to have a dig there every two years instead of annually, as we had done up to that time. The logistics of a full-scale three-month dig had reached the stage where we needed to devote nearly a full year in order to make the dig happen. Strictly speaking, after we made this decision, we have since only managed to adhere to this schedule once, in 1988 (when there was no dig), although most major efforts at the Cove were two years apart after that until the end in 1994.

After four seasons' work at Dinosaur Cove, the balance of 1987 and much of 1988 was devoted to analyzing and writing up the results of that work for scientific and popular publications. Together with a number of colleagues, we wrote a short summary for the professional journal *Science,* which listed the various kinds of vertebrates known to be present in what was polar southeastern Australia between 100 and 125 million years ago and then considered a number of questions about this biota, such as how its members coped with the problems of survival at a paleolatitude inside the Antarctic Circle of the day.[18] This problem had to be addressed, for although it may not have been as frigid then as such latitudes are today, it was certainly not particularly warm, and the area would have been subjected to prolonged periods of continuous darkness each winter.

A longer technical article in *National Geographic Research* covered much the same ground in its conclusions but was devoted primarily to describing what was known about the hypsilophodontid dinosaurs.[19] We named two new hypsilophodontids, *Leaellynasaura amicagraphica* and *Atlascopcosaurus loadsi,* and recognized three more, one of which had previously been named *Fulgurotherium australe* based on a specimen from the opal field at Lightning Ridge, New South Wales. The other two hypsilophodontids were represented by fossil specimens complete enough to strongly suggest that each belonged to a genus and species other than the three for which names could be given. However, they were so fragmentary that they could not be confidently placed in any previously known species from elsewhere nor is it likely that any fossils found in the future could be recognized as belonging to the same species as either of them. For this reason, no species name was given to them nor were they adequate to serve as the name bearers or holotypes of new species.

Figure 25. Leaellyn holding part of the skull of the dinosaur named for her, *Leaellynasaura amicagraphica*.
Photographer: Peter Menzel.

● The Hypsilophodontidae are small, generalized, bipedal ornithischian dinosaurs with an appearance superficially similar to that of a kangaroo. Like kangaroos, they had long tails, small heads, reduced forelimbs, and herbivorous jaws. Unlike the kangaroos, to which they are no more related than are other dinosaurs, they probably ran rather than hopped, did not have external ears or fur, and laid eggs rather than carried their young in a pouch. They appear in the Jurassic, having evolved from a somewhat more primitive group, the fabrosaurids, which are rather similar in overall appearance and are among the earliest of dinosaurs. Although their fossil remains have been found on all continents, they are typically, but not always, quite a rare element in the dinosaur assemblages where they occur. This may in part be because of their small size. It is widely held that the more specialized ornithischians, such as iguanodontids, ceratopsians, pachycephalosaurs, and hadrosaurs, evolved from ancestors structurally close to the hypsilophodontids, if not from the hypsilophodontids themselves. Although they gave rise to these groups, the hypsilophodontids themselves persisted as a distinct group until the end of the Cretaceous. Strangely enough, the last hypsilophodontid known, *Thescelosaurus neglectus,* is structurally the most primitive member known of the family. This suggests that there were many as-yet-undiscovered kinds of hypsilophodontids.

What was particularly remarkable about our hypsilophodontids was that there were so many of them. Elsewhere in the world, this group of dinosaurs, if it is present at all, is generally a rare part of the fauna. Yet in polar southeastern Australia, they were more than half of the dinosaurs we found and formed a significant part of the biomass. Although there is a clearly pronounced size bias in favor of small bones in the southeastern Australian dinosaur record, that alone cannot explain the dominance of the hypsilophodontids there. At minimum, there are three hypsilophodontid genera in Victoria. By comparison, only two hypsilophodontid genera have been found in the Morrison Formation of the western United States. In total, about 100 different dinosaur genera have been recorded from that rock unit. Such low diversity is typical of most rock units that have yielded any hypsilophodontids at all.

The remarkable preservation of the endocast of the top of the brain of *Leaellynasaura amicagraphica* provided a clue that this dinosaur might have had unusual visual acuity. The optic lobes of its brain were larger than those of the four hypsilophodontid specimens from lower latitudes with an exposed endocast of the brain. Because they lived at polar latitudes and had this attribute, the most plausible reason for the difference appears to have been enhanced visual ability, an advantage for an animal active during the winter in an area with prolonged periods of continuous darkness.

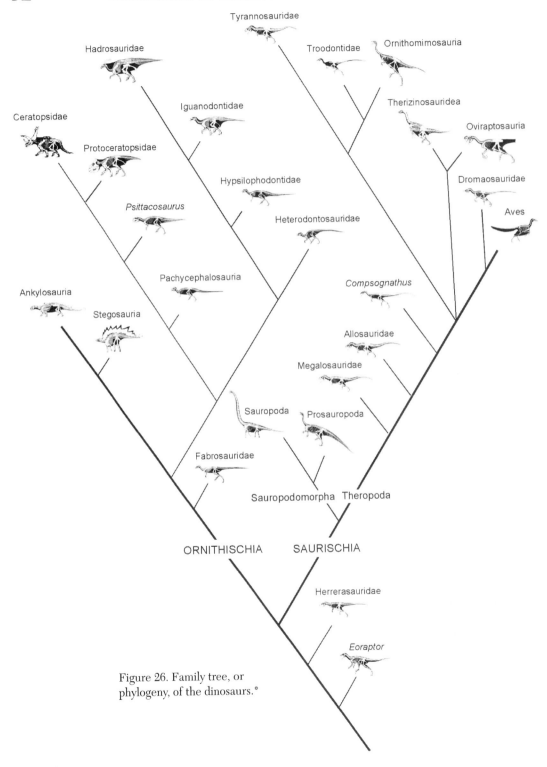

Figure 26. Family tree, or phylogeny, of the dinosaurs.°

°After Currie & Padian (1997), Novas (1997), and Paul (1988, 1997).

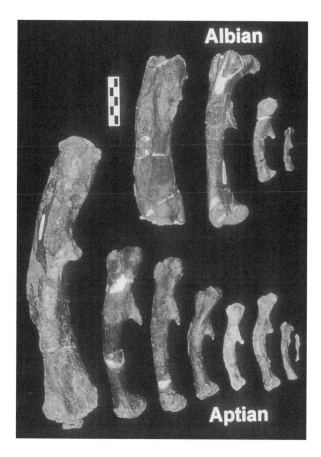

Figure 27. Victorian hypsilophodontid femora. The upper row of femora are Early Albian in age. The Albian is a stage of the Cretaceous Period that extends from 112 to 97 million years ago. The lower row are Early Aptian, a stage which extends from 124½ to 112 million years ago. Although where to draw boundaries is not obvious, the extent of the variation seen in these femora strongly suggests they represent perhaps as many as half a dozen genera.

Atlascopcosaurus loadsi was named in honor of the Atlas Copco organization, without whose mining machinery the fossils would still be in the ground, and for their Victorian manager, Bill Loads, who organized this support. The derivation of *Leaellynasaura amicagraphica* is more complex. Our daughter Leaellyn wanted a pet dinosaur of her very own when she was two years old. She thought Christmas would be a good time for her to get it. This seemed a quite reasonable request to her because, with the impeccable logic of a two-year-old, she knew that her parents were searching for dinosaurs, and it only seemed fair that she should have one too. We patiently explained to her that living dinosaurs were not what we were after, and so the only dinosaur that we could promise her for her very own was one that would have her name. In the end, she had to wait ten years until it was found and officially named.[20] The root *amica* means "friend" and honors the Friends of the Museum of Victoria, without whose unswayable prodding it is likely that the systematic excavation at Dinosaur Cove might never have gotten started. And finally, *graphica* refers to the ever-patient, always supportive National Geographic Society for their long-term faith in and backing for our work.

⊛ A companion article in the same issue of *National Geographic Research* by two associates of the project, Barbara Wagstaff and Jennifer McEwan-Mason, described in detail the basis for the age assignments for all the sites where fossil terrestrial vertebrates had been found in the Otway and Strzelecki groups.[21] This dating was done using various kinds of fossil pollen and spores recovered from rocks associated with the dinosaurs. Fossil pollen has the advantage over dinosaurs for dating rocks because it is much more common. While it has been estimated that there are perhaps 2,000 dinosaur skeletons in all the museums in all the world, one can easily recover far more than 2,000 fossil pollen grains from a single gram of rock in the Otway Range!

Dating Methods

How do we know how old a fossil is? Until Henri Becquerel discovered radioactivity in 1896, there was no way of answering that question in years. Prior to that discovery, fossils could only be dated relative to one another. That is, Fossil A is older than Fossil B because A always occurs in rocks beneath those in which B is found. Building on such information, time units were defined by their fossil content in certain rocks. In this way terms such as Jurassic and Cretaceous were proposed for rocks and their fossils respectively in the Jura Mountains of Switzerland and the chalk of the London-Paris basin. Although such units are not based on physical time, they can be utilized to order events in geological history in much the same way that Chinese history can be ordered by a succession of dynasties.

No matter what temperature or pressure a radioactive isotope is subjected to on Earth, its rate of decay remains the same. Because of this, the age when a radioactive rock was formed can be measured in a matter analogous to telling time with an hourglass. The undecayed radioactive atoms correspond to the sand in the top of the glass, while the daughter products they decay into correspond to the sand below. The ratio of one to the other gives the age of the rock.

There are many different radioactive isotopes which have been used for dating. One of the best known of these relies on C^{14}, or carbon 14. Half of a sample of C^{14} decays into N^{14}, or nitrogen 14, in 5,730 years. This period of time is called the half-life of C^{14}. Because of its rate of decay, carbon 14 is very useful for dating items up to 35,000 years old; with the most recent techniques, dates up to 70,000 years are possible. But older rocks and fossils must be dated using isotopes that decay much more slowly. Some of these long-lived isotopes are those of uranium and potassium, whose half-lives are measured in billions of years.

Rocks and fossils are dated today using both fossils and isotope analysis. We dated the Victorian dinosaurs with both methods.

The method that gives the greatest relative accuracy is one that uses pollen and spore fossils collected at the sites of interest. These fossils have been compared with fossil pollen and spores in other parts of Australia where they occur together with the fossils of planktonic marine micro-organisms. Those planktonic micro-organisms are found elsewhere in rocks that can be radiometrically dated. Thus, a chain of relationships, or correlations, enables us to use these fossils to give ages in years for the dinosaurs. The fossils also give ages defined directly by the fossils in the same way that the Cretaceous originally was recognized as outlined in the first paragraph of this section (Dating Methods). While the units are recognized in much the same way, they span shorter periods of time and thus are more precise; for example, analogous to the reigns of individual emperors within a Chinese dynasty.

The particular radiometric method used for dating our sites is called fission-track dating. It relies upon the fact that in any crystal, such as a crystal of zircon, there is always a trace amount of uranium present. The zircons in the rocks where the dinosaurs are found in Victoria were formed as volcanic rocks solidified. Once the crystal had cooled below what is called its annealing temperature, 200°C, any uranium that subsequently decayed caused a flaw in the microstructure of the crystal. When a uranium atom breaks into two smaller atoms, their nuclei strongly repel each other electrostatically. Consequently they fly off at high speed in opposite directions, knocking about the atoms in the zircon crystal lattice, causing a flaw. The longer it has been since the crystal cooled, the more uranium has decayed. Thus, by counting the number of flaws, one can measure the age of the crystal. The flaws are like the sand grains in the bottom of the hourglass. To know how much sand is in the top of the hourglass, one needs to measure the amount of uranium in the crystal. This is done by first heating the crystal after the flaws in it have been counted. The heating takes the crystal above the annealing temperature so that all the flaws in it are erased. The now nearly flawless crystal is subjected to bombardment by a known quantity of neutrons. This causes some of the remaining uranium to fission immediately. By counting the newly formed flaws, one thus estimates, as it were, the quantity of sand in the top of the hourglass. The ratio of the two gives the age.

While fission-track dating does not give as accurate an answer as dating using fossil pollen and spores that have other radiometric dates associated with them, the dates using these two methods do overlap.

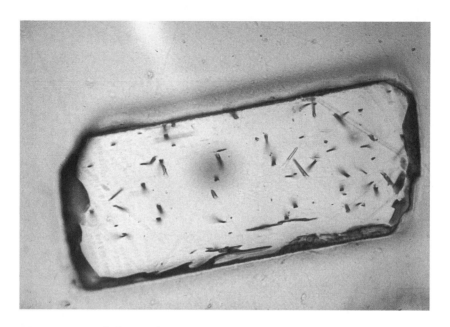

Figure 28. Magnified view of a zircon crystal with fission tracks etched out in it. Fission tracks in the crystal are produced when decaying radioactive atoms emit particles that damage the crystal structure. By etching the crystal with hydrofluoric acid, the damaged zone caused by the flight of atomic particles through the crystal lattice is enlarged. The density of those tracks is directly related to the length of time decay has been occurring. By counting the number of tracks in the crystal, its age can be determined. (I. Duddy)

This corroboration is important for raising confidence in the pollen date, ensuring that there is not some unsuspected systematic bias giving a highly erroneous result.

We are trained as biologists and geologists with a specialized knowledge of paleontology. As such, we cannot know all the latest ideas in all the fields of science that are relevant to a broad understanding of the dinosaurs of southeastern Australia. Of necessity, in order to acquire that knowledge, we rely on specialists in other fields. There are so many ways of looking at our research area that while we ourselves have often identified the fields of expertise that we needed, in other cases experts have come forward on their own initiative to contribute useful information.

In the early phase of the work at Dinosaur Cove, we knew from our preliminary study that when the dinosaurs we found there were living animals the area had been located at high paleolatitude.[22] However, we had no idea how to go about estimating what the mean annual temperature was. For all we knew, although the area was then at polar latitudes, the climate could have been temperate. In fact, that is what had been previously proposed.[23] One of the first persons to see an overlap of interests was Bob Gregory, and we learned that he had a way of making such a temperature estimate. He was then at Monash University in the same department as Pat.[24] He is a geochemist with a particular interest in measuring the ratio of stable isotopes in a geological context and interpreting their significance.

Figure 29. Robert Gregory. *Photographer: Peter Menzel.*

Two of the isotopes that Bob was particularly interested in were oxygen 16 (O^{16}) and oxygen 18 (O^{18}). Because both isotopes have the same number of protons in the nucleus (8), they are chemically identical. However, because O^{18} has two more neutrons, and therefore more mass, it reacts just a little more slowly than O^{16}. As the temperature of various physical and chemical reactions rises, the difference in the reaction rates increases. This results in differing amounts of O^{16} and O^{18} reacting. It is because of that temperature-related difference that a method to estimate mean annual temperatures could be developed.

Concretions are spherical masses of rock that are formed in sediments because of cementation that starts around a nucleus such as an individual crystal, sand grain, or even a fossil. The cement is commonly calcium carbonate, the primary constituent of chalk. In the rocks of the Otway and Strzelecki groups, such concretions were formed in the loose sediments before they had fully hardened into rocks. The calcium carbonate cement was deposited by groundwater that flowed through the still-loose sediments. The ratio of O^{16} to O^{18} in the carbonate ions dissolved in that groundwater reflected the mean annual temperature of the surrounding area. Those carbonate ions eventually became part of the calcium carbonate that cemented the concretion together. By measuring the oxygen isotopic ratio, Bob had a way of estimating the mean annual temperature that prevailed soon after the sediments were laid down but before they had completely hardened into solid rock by being deeply buried in a subsiding basin.

The value that Bob came up with for the mean annual temperature surprised us. He estimated that the mean annual temperature was somewhere near –2°C (28°F).[25] Minus 3°C (27°F) is the mean annual temperature of modern-day Fairbanks, Alaska. That value seemed incredibly low to us, considering the diverse flora that existed then as well as the vertebrates that lived along with them. We, therefore, actively encouraged the work of others who were attempting to measure the mean annual paleotemperature utilizing quite different approaches to the problem. The question of what precisely the temperatures were at particular times is still open, but Bob's work provided us with a starting point, whereas before we had no inkling at all about what the paleotemperature might have been.

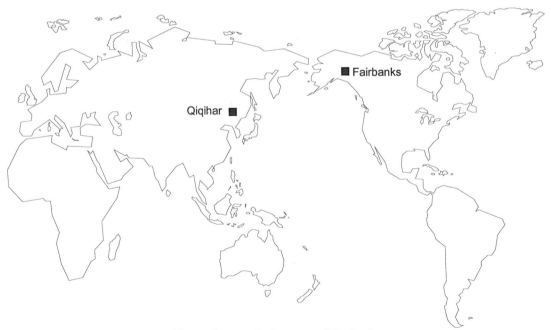

Figure 30. World map showing the locations of Fairbanks, Alaska, and Qiqihar, China.

5

Underground at Dinosaur Cove

The article in *National Geographic Research* appeared in the same month that digging started anew at Dinosaur Cove—January 1989. Over the years, a great quantity of gear had been accumulated to carry out the job. The gear was stored partly in Melbourne and partly in a number of caravans (mobile homes) that each winter were hauled to the Department of Conservation and Natural Resources depot in Beech Forest. By 1989, gathering all this equipment together at the beginning of a dig and dispersing it at the end was becoming a major project that fell particularly heavily on John Herman.

The Temple of Janus

Storing caravans outside at Beech Forest was not ideal, because for much of the year heavy mist blows over the ridge on which the town is located, promoting rust. The first caravan acquired by the project was provided by the Friends of the Museum of Victoria and dubbed "The Dinovan" because of a dinosaur Pat had painted on its side. As it was towed back to Dinosaur Cove at the beginning of the 1989 season, the rim of a wheel collapsed, having finally rusted completely through.

John Herman and Tom resolved then and there that a shed needed to be built to store the caravans out of the weather. John offered to provide the materials if Tom could divert the labor necessary to construct it. This was at the beginning of the field season, and there was not time to recruit many additional people for an unexpected labor-intensive project. However, the Fates had decreed that there would be an excess of volunteers about a month and a half after the work began. Stephen Seavey of Lewis & Clark College, Oregon, U.S.A., had arranged to bring a group of twenty-five students to Dinosaur Cove for a week. In addition, an army unit would be on site before that for another week.

The first thing needed for such a shed was a place to put it. David and Winsome Denney, on whose land we had camped every year since the project began, graciously agreed to let us build it on their land. As John brought down other gear to Dinosaur Cove over the next month and a half, he also brought the raw materials to construct the shed. Had all the supplies necessary to construct the shed been on site when the group of Australian Army volunteers from the 21st Construction Squadron at Puckapunyal arrived in early February, the five of them could probably have built it in the time they were there. As it was, they set the stumps (foundation) in a professional manner before they left.

Directed by Tom Whitelaw, brother of Michael Whitelaw, several of the students accompanying Steven Seavey turned to with a will and constructed the shed quickly. In places it was 15 centimeters (6 inches) out of plumb, but it was up. After the departure of the Oregon students, Gordon Spark, an experienced carpenter, single-handedly managed to correct many of the alignment problems. It was sheltered by trees, had solid walls built out of reinforced timber from machinery packing cases, and was covered with corrugated iron. It was quite sturdy and did not move even in the highest winds. There was enough space to store two caravans, and the bulk of the equipment used for the dig itself could be kept there as well. And not another caravan wheel collapsed. While this was going on, the work of excavating fossils went forward in Dinosaur Cove without a hitch caused by an insufficient number of people.

When a storage facility such as this is built, one naturally assumes that it will have doors. However, it soon became apparent that putting in doors large enough to provide access for the caravans was going to be both difficult and costly. A much easier solution was taken. Because there was no need to close up the shed when the field season was under way, there was really only a need to open the shed at the beginning of each annual excavation period and close it at the end. Therefore, instead of building an elaborate hinging or roller device, the door was simply nailed onto the 'shed at the end of the season and pulled off at the beginning of the following one. Because the shed bore an analogy with the Roman temple of Janus, which was opened in time of war and closed in time of peace (a rare event in Rome), we christened it the "Temple of Janus."

Another problem that plagued work at Dinosaur Cove from the beginning was a lack of horizontal space within the cove to store equipment out of reach of the waves. Near the entrances to the Slippery Rock site there were no convenient ledges, and gear sometimes fell between boulders and became difficult to retrieve, if not completely lost.

Graeme O'Brien, a volunteer who had answered the 1986 advertisement in *The Age* newspaper, resolved to do something about this. He had experience constructing scaffolding while working on the Westgate Freeway in Melbourne and elsewhere, and he built with the help of several other volunteers a platform about 2 meters (2 yards) higher than the entrance to the tunnels at Slippery Rock. When we saw it go up, seemingly precariously clinging to the rocks, and thought of the large boulders that can be moved in high seas across the shore platforms, we were uncertain how long such a structure, useful though it would certainly be, would last. In the end, it was never damaged by the sea nor was anything ever lost from it, although there were many occasions when angry waters swirled around its base.

Figure 31. The spidery-legged outside platform clinging to the rock face just to the left of the openings to the West and East Tunnels at the Slippery Rock site. *Photographer: Peter Menzel.*

Boggers and Barrows

The 1989 excavations were to last for eleven weeks, nearly a month longer than any previous dig at Dinosaur Cove. In that time, a number of people acted as the mine manager, because no one person could give it their undivided attention. Principal among these were Rebecca Norton and Michael Charlesworth, both thorough professionals but two

Figure 32. Rebecca Norton.
Photographer: Peter Menzel.

quite different people. Whereas Rebecca was a miner first, last, and always, Michael added the skills of an excellent fossil spotter to his list of accomplishments.

Seeing a situation developing where there might be no mining manager available for prolonged periods because no qualified individual was able to commit their time totally to the project, Pat O'Neill made the decision to officially certify Tom as a mine manager. The mine manager's license issued to him was probably the most restricted one ever granted in the state of Victoria. The important thing was that under its terms, when there was no blasting going on, work could proceed legally underground when he was present. The most important qualification of a mine manager is a proper sense of safety. Because he had observed him over five years, it was Pat's professional opinion that Tom was sufficiently cautious to get people out of a situation before it became unduly dangerous, so he issued the license.

Almost all effort in 1989 was focused on the Slippery Rock site. The twin goals there were to widen the First Cross Tunnel in anticipation of obtaining more fossils immediately adjacent to the area which had produced so much in 1987, and driving farther underground in narrow tunnels to see if the fossiliferous unit extended significantly north or east.

Figure 33A. Underground at Dinosaur Cove. (See also Plate 9.) The woman operating the jackhammer, Helen Wilson, is standing just below the level of the fossiliferous layer. It lies about ⅓ meter or 12" below the level of the rock into which she is digging. Having established by careful examination that the rock above the fossiliferous layer was almost certainly devoid of fossil bone, it was blasted away in order that she could work down to the fossiliferous layer. Compared with digging horizontally, working downward is less likely to damage the fossils. *Photographer: Peter Menzel.*

Figure 33B. As the fossil layer was exposed, it was first carefully examined. If a fossil was seen at that stage, cuts were made in the surrounding rock with a rocksaw. Chisels were then inserted into those cuts and hit with a large hammer, breaking free the fossil in a rectangular piece of rock.

Figure 33C. The vast bulk of the fossiliferous layer had no fossil bone showing on the surface. However, rock which might well have a fossil in it and which had been so hard to win was not thrown away because nothing was showing on the outside. Rather, this rock was taken to volunteers who patiently made big rocks into little rocks. They examined each fresh surface as they broke up the rocks with hammers and chisels. At any one time, about half the crew would be carrying out this vital task, finding one or two fossil bits of bone on a typical day, most of which would be of little scientific value. But once in a while, something spectacular turned up, and that was the reason they kept at it. *Photographer: Steve Morton.*

The East and West Tunnels were driven forward so that by the end of the dig, their farthest points of penetration were 15 and 18 meters (16 and 20 yards) from their respective entrances. It had been planned to link those deepest ends with a Second Cross Tunnel in 1989, but progress was so slow that it was not to happen for another year. Lack of time also made it impossible to dig up the floors of the East and West Tunnels to see if fossiliferous rock might be found in one or both of them beyond the First Cross Tunnel. That, too, would have to wait for another year.

What was possible in the time available was to take small-diameter cores at the backs of both the East and West Tunnels to see if the fossiliferous layer might be present in either of them. Rob Anderson of the Melbourne Metropolitan Board of Works had coring equipment available to him and volunteered to drill the cores. The results of his efforts did not portend well, for nothing was found in either the East or West Tunnel.

A few volunteers continued to make progress at Dinosaur Cove East, recovering a few hundred bone fragments over the field season there. In order to assess the likelihood of that fossiliferous unit continuing northward, Rob Anderson took cores there as well, and the result was a clear indication that the dinosaur-bearing layer extended 5 meters (16 feet) beyond the area that had been excavated up to that time. On that basis, explosives were used to remove the overburden in the newly identified fossiliferous area in preparation for digging there in 1991.

In order to speed up the rate of tunneling progress, a "bogger" was employed (see Fig. 21E). Drawn by a steel cable pulled by a pneumatic winch, a bogger is a steel scoop that is used to pick up rock fragments and transport them elsewhere. At times these devices worked well to move the rock out of the tunnels that had been broken up by blasting, but often, because of mechanical problems with the pneumatic winch, the rate of progress when we used them was no greater than when the volunteers with shovels filled wheelbarrows to muck out the tunnels after each blast. Steel cables under tension that were being operated for the most part by people who were inexperienced in their use were flying around; there was always a nagging fear in the back of Tom's mind that a serious accident would take place. Fortunately, one never happened. Wheelbarrows may be low tech, but are a lot safer in the hands of people who have never worked underground in their lives, and using them made Tom feel much more comfortable, even when he was physically worn out by having done his share of the shoveling.

The East Tunnel had been located near the middle of the original exposure of fossiliferous rock at the Slippery Rock site. To the east of it, the cliffs were and are unstable because of deep cracks and prominent overhangs. The only way to reach this area in relative safety seemed to be to tunnel eastward out from the East Tunnel, starting at a point well inside from the entrance. Joints in the rock and patches of unstable rock underground made advancing eastward slow, nerve-wracking work. After pushing forward 3 meters (10 feet) in this direction, Rebecca Norton suggested that it would be prudent to stop, which we did immediately.

If in the future more fossils are sought from the Slippery Rock site, the area east of the East Tunnel would be the most likely place to try. The best way to do it might be to monitor the locality in the decades and centuries to come. As the relentless pounding by the sea causes the cliff to disintegrate, blocks of rock containing the known fossiliferous layer will break free from time to time and can then be collected in relative safety.

The First Cross Tunnel was widened twice by blasting its north wall, removing about one meter (one yard) each time. After the first drag-cut (as this type of blasting is called), the fossiliferous layer continued to produce numerous fossils over almost the entire distance between the East and West Tunnels. As was initially the case, the thickest and most prolific area was at the western end of the First Cross Tunnel.

After the second drag-cut in the First Cross Tunnel, we found that the fossiliferous layer was confined to the westernmost one-third of that tunnel and that it was thinner than before. In light of these results, it was now clear that the best areas at the Slippery Rock site for fossils were the original outcrops where fossils were found and the southern side of the First Cross Tunnel. This suggested that at some stage the Pillar of stone that had been left behind to hold up the roof between the entrances of the East and West Tunnels should be knocked down and searched for excellent fossils like those that had been collected on its northern and southern flanks.

Rebecca Norton maintained that the Pillar was a basket case. She contended that it was full of fractures that had shrunk and pulled away from the overlying rock and that it therefore provided no support to the roof in any case. However, Tom was not confident enough to simply dig it away with the assumption that the roof would hold. In order to hold the basket case together, Rebecca had supervised the insertion of numerous steel split sets, which are essentially giant rock nails, into it. The split sets did double duty by also holding wire mesh to the sides of the Pillar. These measures held the rock together in what was, in effect, a giant basket. Certainly if the roof had started to move, this basket would have kept it from falling to the floor and crushing all who were in its way.

But the idea was now in our minds that sooner or later the Pillar would have to go. Half in dread, Pat O'Neill had earlier predicted this as an eventual course of action soon after the creation of the Pillar when fossils turned up next to it in abundance in the First Cross Tunnel.

When the work was finished at the end of March, there were no obvious prizes such as the partial skeleton of *Leaellynasaura amicagraphica* found in 1987, although there were numerous bones of high quality, so the season was far from a failure. In the months to come Lesley Kool would systematically go through the blocks of rock collected, beginning with the most promising, and expose the fossils in order to determine what was there.

One specimen that had turned up was a tibia and fibula together with the astragalus, the bones of the lower leg and ankle. What was remarkable about it was that the tibia had obviously been diseased. What tipped us off was that additional bone, irregular in form, had grown around the original shaft of healthy bone. In our ignorance, we tentatively interpreted this as the result of the bone having been broken and the irregular bone having been laid down as a pathological response to that condition. Ultimately, we had the good sense to approach Janet Gross, a pathologist at the University of Melbourne, who immediately recognized the symptoms of osteomyelitis, bone deterioration brought about by a chronic infection, the oldest instance of this condition on record.

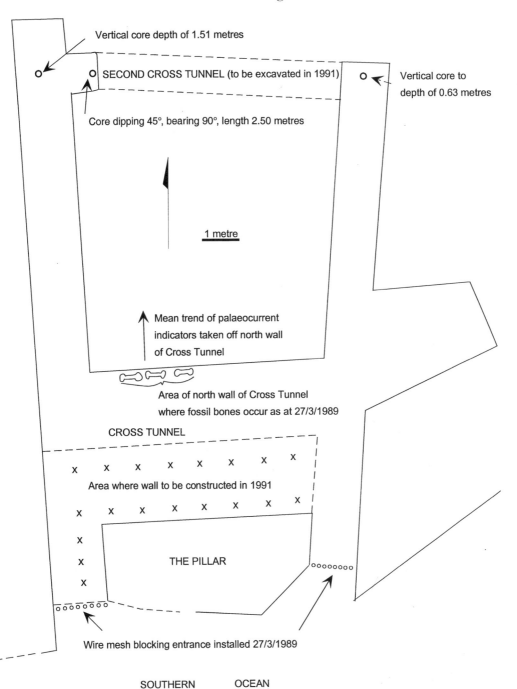

Figure 34. Map of the tunnels at the Slippery Rock site at the end of the 1989 excavations.

While we were overseas trying to identify more of the fossils during the southern winter of 1989, Lesley Kool continued her work, preparing the blocks of rock. When we returned to Melbourne, she had a surprise waiting for us. Calling us into her laboratory, she showed us a black-draped object which, when the velvet covering was ceremoniously removed, turned out to be the partial skeleton of yet another hypsilophodontid dinosaur, found less than 2 meters (2 yards) from where *Leaellynasaura amicagraphica* had been located. Lesley had started to prepare a block of rock found next to where the pathological specimen had been collected. Initially only a few fragments of bone were showing on the surface, but as she dug in deeper and deeper, more bones kept appearing until she had most of a hind leg, a pelvis, and a string of vertebrae. She had found more of the same individual of which the pathological specimen was a part. Lesley's discovery meant that when we described and analyzed this pathological specimen, we knew what a normal tibia looked like.[26] Because of the differential growth between the two bones, we could establish that the condition had persisted long enough for the difference in their lengths to have developed.

This was just one of many times when Lesley pleasantly surprised us by her seemingly magical ability to turn what in the field had appeared to be a specimen of little consequence into a highly significant find.

Just a Little Dig

Because we did not intend to have another dig at Dinosaur Cove for two years, shortly after the 1989 effort was over Tom started planning for 1991. It soon became apparent that the failure in 1989 to excavate the Second Cross Tunnel and dig up the floors of the East and West Tunnels inward of the First Cross Tunnel was going to make planning difficult.

If there were rich fossiliferous rocks in those as-yet-unsampled areas, then there was no point in going to the trouble of removing the Pillar and collecting what fossiliferous rock might be under it. On the other hand, if those unsampled areas proved sterile or only marginal, then a big commitment was going to have to be made well in advance of the start of the 1991 dig in order to start the process of excavating the Pillar that field season. A decision to begin the first phase of the process to remove the Pillar could not be taken and then implemented halfway through the dig. This was because the initial step would be to pour a concrete pillar alongside the Pillar in order to support the roof when the Pillar was removed. If this was to be done during the 1991 dig, organizing the logistics to do so would have to begin months before the start of that excavation. To make this decision whether or not to go ahead with removing the Pillar, another month of excavating would be

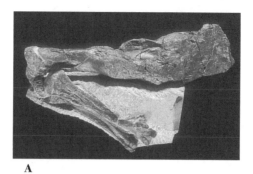

A

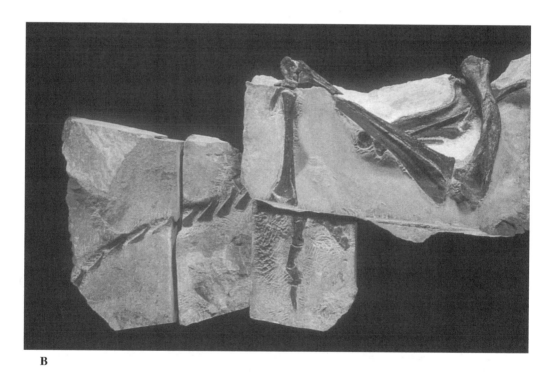

B

Figure 35. A. Left metatarsals, astragalus, and pathological tibia of a hypsilophodontid dinosaur recovered from the First Cross Tunnel at the Slippery Rock site,
1989. B. Pelvis, right hind limb and posterior vertebrae of the same individual as in
Fig. 35A. It now appears quite likely that this specimen is another individual of
Leaellynasaura amicagraphica.

needed. January to early February 1990 was the time chosen to do this work. Because there was no intention to collect large quantities of fossiliferous rock, the flying fox was not installed. This meant that all the equipment had to be carried in on people's backs, but that was far less total effort than what installation and removal of the flying fox would have required.

For the excavating of the Second Cross Tunnel, we had the assistance of Ian Jesser, an experienced miner who worked well with our crew of amateur volunteers. Like all who had worked at the Slippery Rock site before him, his first few blasts failed to behave the way he expected, but eventually he developed a pattern that worked successfully. In six days, he drove the 6 meters (20 feet) between the ends of the East and West Tunnels and shortly thereafter departed.

The balance of the time was spent taking up the floors of the Second Cross Tunnel and the East and West Tunnels inward of the First Cross Tunnel. Nothing was found in the East Tunnel and only a narrow channel 70 centimeters (2 feet 4 inches) wide with a maximum thickness of 3 centimeters (1 inch) was found in the Second Cross Tunnel. In the West Tunnel, however, fossiliferous rock occurred over almost its entire length, starting from a feather edge only 30 centimeters (1 foot) from where Rob Anderson had taken a core the year before and thickening to a maximum of 8 centimeters (3 inches) near the First Cross Tunnel. Only a few fossils were found in it, but the rock consisted of chunks of clay in a sandstone resting on a massive mudstone and so was similar to the fossiliferous rock which had been so productive in the First Cross Tunnel. The question that remained unanswered was the relationship between the two similar channel deposits. Two possibilities suggested themselves. First, this newly discovered fossiliferous rock was continuous with the fossiliferous rock that we already knew about. Second, the two rock units represented two separate episodes of stream deposition, perhaps separated in time by only decades, centuries, or millennia.

If one flies today over an area where billabongs or oxbows occur, what is commonly seen is that they form arcs that describe about one-quarter or one-third of a complete circle. These represent channels that have been cut off as an active stream meanders back and forth across the landscape. Frequently, such arc-shaped ponds lie close to one another. Where they are closest together, if one is perpendicular to a second and cuts across it, it is evident that the two were active channels at different times.

It was important to work out if this newly found channel deposit was part of the first one originally found. If so, we could expect it to have fossil bones in it because the first one discovered did. However, if it was formed at a different time, it could well be that the conditions were different and that there just happened to be fewer bones lying around

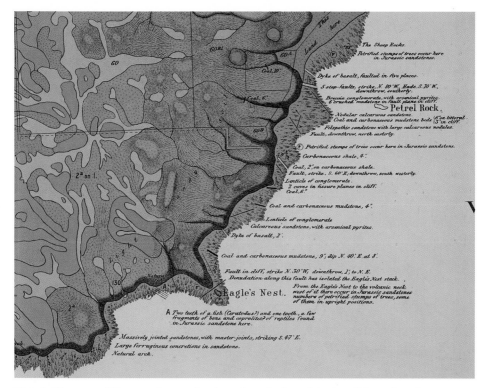

Plate 1. The part of Ferguson's map showing where he found the lungfish tooth and toe bone of *Megalosaurus*.

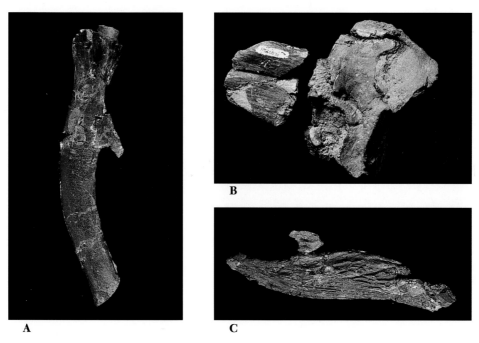

B

A

C

Plate 2. Three of the first thirty bones recovered from the coastline between San Remo and Inverloch in 1978 by Tim Flannery. A. Femur (thigh bone) of a small ornithischian ("bird-hipped") dinosaur. The prominent spike or 4th trochanter in the middle of this specimen immediately identified it as the first undoubted dinosaur recovered in 1978. B. Astragalus or ankle bone of an allosaurid, a carnivorous dinosaur, perhaps *Allosaurus* itself. C. The mystery jaw fragment that eventually proved to belong, quite unexpectedly, to a labyrinthodont amphibian, a group that in 1978 was thought to have become extinct more than 80 million years before this individual lived. Nineteen years later, it would finally be officially named *Koolasuchus cleelandi* (Warren, Rich, & Vickers-Rich 1997). *Photographer: S. Morton.*

Plate 3. Restoration of the Early Jurassic labyrinthodont *Siderops kehli* from Queensland, Australia (*artist Frank Knight*). Known from nearly a complete skeleton, when this animal was named and described in 1983, it was the youngest labyrinthodont known (Warren & Hutchinson 1983). While *Koolasuchus cleelandi* is closely related to *S. kehli* and looked a bit like it, it is 90 million years younger.

Plate 4. Excavating in the channel deposit at Dinosaur Cove in 1981 uncovered half a dozen additional bone fragments. The three persons shown are working at the then known margins of the fossiliferous unit.

Plate 5. Surf Life Saving Victoria helicopter delivering a load of logs into Dinosaur Cove, February 1984.

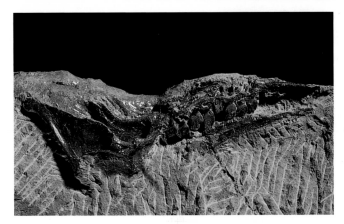

Plate 6. Skull fragment found on the fifty-first day of digging at Dinosaur Cove in 1987. It is now the holotype [= specimen on which a formal scientific name is based] of the small hypsilophodontid dinosaur *Leaellynasaura amicagraphica*. Inside view. *Photographer: F. Coffa.*

Plate 7. Restoration of a group of *Leaellynasaura amicagraphica* moving through undergrowth in winter. In the upper left-hand corner is illustrated a single individual of *Atlascopcosaurus loadsi*, another hypsilophodontid known only from its dentition. *Artist: Peter Schouten.*

Plate 8. The understory in the Otway Range today. A very similar lush flora existed 106 million years ago a few kilometers away from where this picture was taken at what is now Dinosaur Cove. *Photographer: Peter Menzel.*

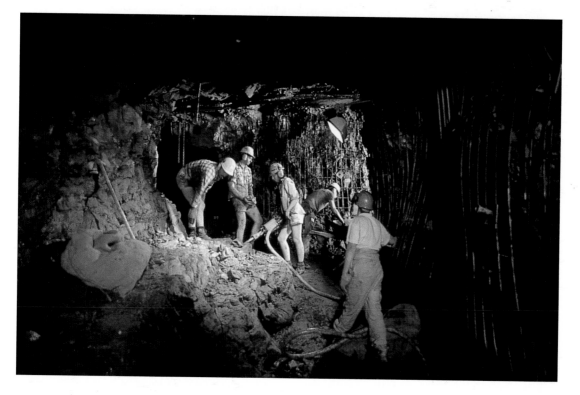

Plate 9. Underground at Dinosaur Cove. *Photographer: Peter Menzel.*

Plate 10. Restoration of *Timimus hermani. Artist: Peter Trusler.* Courtesy of Australia Post.

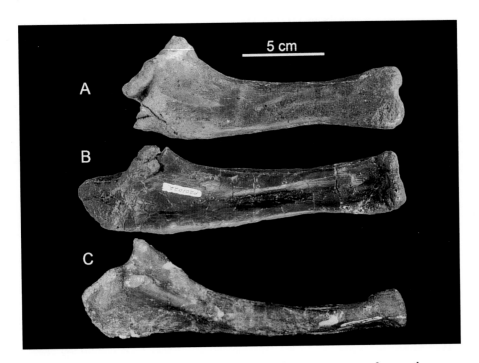

Plate 11. A. Left ulna, or forearm bone, of a presumed protoceratopsian from rocks about 115 million years old at The Arch near Kilcunda. B. Left ulna of *Leptoceratops gracilis* from rocks about 67 million years old near Drumheller, Alberta, Canada. C. Ulna of a carnivorous dinosaur from rocks about 106 million years old from the Cross Tunnel at Dinosaur Cove. Although carnivorous dinosaurs typically have foreshortened forelimbs, the individual bones have the long, slender proportions typical of many land-living vertebrates. In contrast, among neoceratopsias, the ulna is characteristically short and stout and is flattened from side to side, a pattern unique to that group. *Photographer: S. Morton.*

Plate 12. Restoration of *Leptoceratops gracilis,* an animal probably quite similar to the one whose ulna was found at The Arch. *Artist: Peter Trusler.*

Plate 13. The specimen on which the name *Leaellynasaura amicagraphica* is based, pictured as it might have appeared in the first step on the way to becoming a fossil. *Artist: Peter Trusler.*

Plate 14. Two of the very few pterosaur bones found in the Early Cretaceous of Victoria. The specimen that looks like a mask is two sacral vertebrae, the ones that join the pelvis to the backbone. The rod-like bone is from either the hand or foot. *Photographer: F. Coffa.*

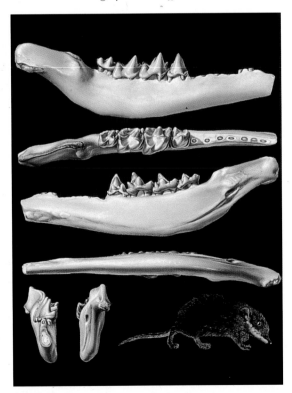

Plate 15. The jaw as restored by Peter Trusler together with a drawing by Draga Gelt of what *Ausktribosphenos nyktos* might have looked like as a living animal.

Plate 16. Vertebra of oviraptorosaur (see figure 65B). *Photographer: S. Morton.*

Figure 36. Aerial view of modern billabongs or oxbows on the North Slope
of Alaska.

in the nearby vicinity that could be swept into the second channel when
it was filled. This second alternative now seems to have been the case.

While the excavation was under way, a number of visits regarding
plans for 1991 were made by people from Western Mining Company
and 21st Construction Squadron of the Australian Army. At the suggestion of John Landy, another person who offered his assistance to the
project, Hugh Morgan, CEO of Western Mining, had approved professional assistance with the tunneling operations by his company for the
next major excavation at Dinosaur Cove. Likewise, the army had expressed interest in constructing the concrete pillar to replace the Pillar.

When the results of the work to date were available, it was decided
to go ahead with constructing the concrete pillar because the meager
nature of the fossils found in the West Tunnel and the almost complete
lack of them in the Second Cross Tunnel and East Tunnel did not au-
gur well. However, enough fossils were found in the West Tunnel to
justify expanding it westward to see if good material could be found
there.

Therefore, the plan was to have Western Mining and the army operate one after the other. Western Mining would excavate to the west of the West Tunnel and then the army would pour the concrete pillar next to the Pillar. Because of scheduling problems, the army requested that the time of their participation be moved forward to December 1990. Therefore, the schedule was for Western Mining to carry out its work during the latter part of November.

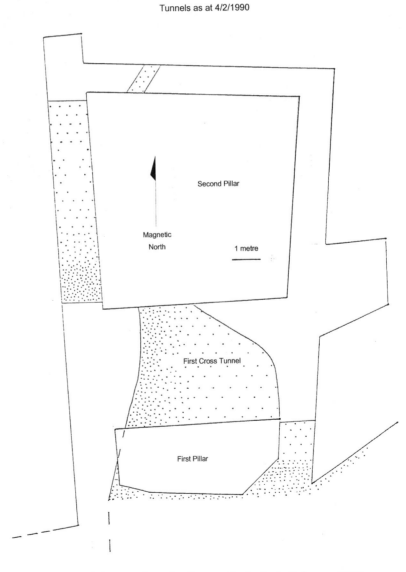

Figure 37. Map of the tunnels at the Slippery Rock site in February 1990.

Life in Dinoville

Crew members who worked all day in Dinosaur Cove returned at the end of the day to a camp made up of a hodgepodge of tents of various sizes, a variety of homemade and professionally manufactured portable buildings, and a few caravans. Most of these were scattered through a woodland formed of trees less than 10 meters (33 feet) tall with an undergrowth dominated by bracken fern. There was an open paddock on one side of the camp and a broad firebreak on the other. Only the caravans and a few tents were in the open. This was because the trees provided a highly effective windbreak. With the Southern Ocean near-by and only open water between the shores of it and Antarctica, a cold, rain-laden wind from the southwest could make life in a tent less than pleasant. Anything that ameliorated those conditions was treasured, and being among the bracken ferns and small trees helped.

But even with the help of the vegetation, life in camp was a battle to stay dry and warm. Tents would flatten in a storm and sleeping bags and clothing would become sodden and invaded with mildew. This could go on day after day. At one stage, things got so bad after a solid week of such conditions that arrangements were made to move the entire crew into a farmhouse for a few days so that they could dry both themselves and their gear out.

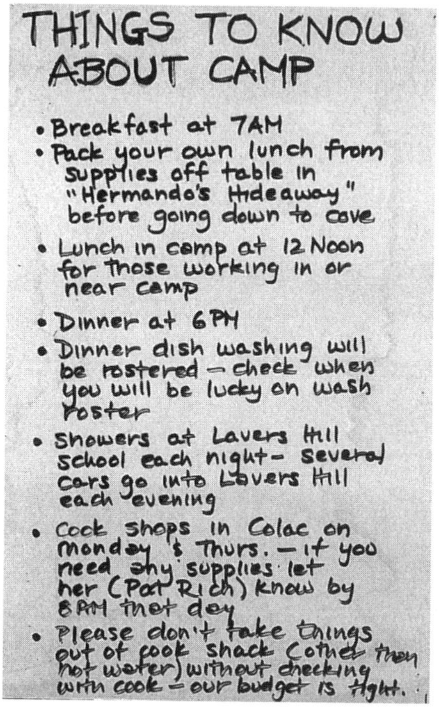

THINGS TO KNOW ABOUT CAMP

• Breakfast at 7AM
• Pack your own lunch from supplies off table in "Hermando's Hideaway" before going down to cove
• Lunch in camp at 12 Noon for those working in or near camp
• Dinner at 6PM
• Dinner dish washing will be rostered — check when you will be lucky on wash roster
• Showers at Lavers Hill school each night — several cars go into Lovers Hill each evening
• Cook shops in Colac on Monday's Thurs. — if you need any supplies let her (Pat Rich) know by 8PM that day
• Please don't take things out of cook shack (other than hot water) without checking with cook — our budget is tight.

Figure 38. Things to know about camp.

Then when the weather shifted around to the northwest, the wind could be a dry blast out of the Centre with temperatures reaching 42°C (108°F). The camp then was a refuge from the furnace-like conditions that could develop in Dinosaur Cove on such days, when the walls on three sides seemed to direct all the heat toward us.

Figure 39. A somewhat modified road sign near Dinoville.
Photographer: Peter Menzel.

Figure 40. Pat drawing water out of the water cart constructed by John Herman.

The social life in camp centered on the recreation tent. It was a massive affair, an umbrella tent that in a pinch could accommodate twenty people or more. It was a very appropriate piece of gear for a paleontological dig because it had been in the family of Bob Tranter, one of the long-term volunteers, for more than fifty years when he loaned it to us. Meals were taken in it, particularly in inclement weather, and in the evening, people tended to congregate there to talk. There was no floor, so in order to keep the dust down, odd pieces of old carpet were acquired to cover the floor.

One year we somehow acquired a large, box-like caravan, which also served as a meeting place for people. It did have the advantage of being a drier place when the rain was almost horizontal.

To supply water to the camp, every few days a trip was made to the Lavers Hill school. There hoses were used to fill three 200-litre (55-gallon) drums that were permanently fixed to a trailer John Herman constructed for this purpose out of odds and ends (as usual). Back at the camp, the trailer was positioned near the cook shack. It was fitted out with a faucet and was thus quite convenient to use.

The Days of November

The Otway coast is a beautiful, rugged place. Much of that beauty is due to the fury of the sea. Often harsh in the winter, at any time of year it can pound the coast relentlessly. This occurs less frequently in the summer, and that is why most of the work at Dinosaur Cove has been done during those months. But to mesh with the schedule of the army, the initial work for the 1991 biannual dig was scheduled to begin in November 1990, late spring.

A few weeks before the work was to start, the army withdrew without explanation. Because events were already scheduled with Western Mining and a pair of volunteers, Graham and Jennifer King (who had quit their jobs to participate beginning in early November), the dig went ahead as scheduled.

Because only a few people were on site, we expected that it would take two weeks to set up the physical plant. Midway through this process, the highest seas in two years swept Dinosaur Cove, floating timbers that had laid undisturbed through two previous winters out of the tunnels at the Slippery Rock site. Other mishaps dogged the work, such as a landslide in the midst of thick scrub on a particularly steep part of the descent into Dinosaur Cove that wiped out 20 meters (65 feet) of telephone wire plus pipes for compressed air and water, and the discovery that a nest of wasps had clogged the water line. Despite these and other, more minor, difficulties, the equipment was in place and operating when David McMahon from Western Mining arrived to begin excavating what would become known as the Western Chamber, a widening of the West Tunnel on its western side between the First Cross Tunnel and the Second Cross Tunnel.

David was an excellent, efficient miner with no fear of underground work, but it was his misfortune day after day to see high seas pounding into Dinosaur Cove. He constantly fought to maintain his composure, for he was brave man having to face a real fear, perhaps an actual phobia, about a turbulent ocean. His skills were needed back at Stawell where he was employed, and so his time with us was limited. It was a measure of his devotion to the project as well as of his bravery that he returned on his weekend off to continue the work as a volunteer.

Secondhand tunnels seldom have much use after their original purpose has been served. However, in the East Tunnel, which had proven utterly barren the previous summer, Michael Marmach, an inventive volunteer, built shelving to store all the gear that previously had been kept outside on scaffolding. Unlike the outside scaffolding, once the shelving was up, it stayed up rather than having to be erected and taken down each season.

During this period, one of the two instances where a human bone was broken at Dinosaur Cove occurred. It did not happen underground where the tunneling was occurring but outside when Graham King was climbing over some rocks. He slipped and fell less than half a meter (20 inches), one leg landing between two rocks as his body continued downward. As a consequence, Graham broke his fibula. Cursing himself for his clumsiness, he managed to hobble most of the distance out of the cove with the help of one or two people. Two-thirds of the way up the track that ascends the 90 meters (300 feet) of vertical distance out of Dinosaur Cove, he was met by sixty-year-old Ray Blanford, who proceeded to sling him over his shoulder and carry him up the rest of the way to where a wheelbarrow was waiting to convey him to a car that took him to Colac and a doctor. His Christmas present to himself was to remove the cast about a month after the incident occurred.

Once the work of excavating the Western Chamber was completed, operations were shut down, to be restarted in January 1991. As planned originally, Graham and Jennifer King remained in camp at Dinosaur Cove to watch over the equipment for the month of December.

The Great Wall

Although only a relatively small effort had gone into the work at Dinosaur Cove East during 1989, the results there were quite satisfactory. In 1991, we resolved to work both that locality and the Slippery Rock site.

Initially, our idea was to use plugs-and-feathers to expose an extensive area at Dinosaur Cove East that was covered with up to 4 or 5 meters (between 13 and 16 feet) of sandstone. The amount of work to accomplish this would have been tremendous. Granting the technical feasibility of clearing the area in this way but questioning the amount of effort involved, Pat O'Neill suggested blasting off the one corner of the target area where explosives could be safely utilized. That could be done in a single shot rather than devoting a full field season to the project, which the original course of action would have required.

Following his suggestion, 6 square meters (65 square feet) of fossiliferous rock was uncovered by a blast carried out by Ian Jesser. Nicholas van Klaveren (who had always been particularly committed to the Dinosaur Cove East site) and others working with him then began systematically to dig out the fossiliferous rock there. Had it not been for their efforts, almost no fossils would have been found at Dinosaur Cove in 1991. As it was, a small but steady stream of isolated bones flowed from that site the entire field season.

Figure 41. Nick van Klaveren in the summer of 1989. *Photographer: Peter Menzel*

One of these was the largest fossil yet recovered from Dinosaur Cove, a femur 43 centimeters (17 inches) long. A nearby specimen only 45% as long proved to be the femur of a juvenile of the same species, soon to be named *Timimus hermani,* the first ornithomimosaur (or ostrich-like dinosaur) from Australia and one of the oldest species of that group.[27] The species name is in honor of the indomitable John Herman and the generic name is for Tim Flannery, co-discoverer of the site with Michael Archer, and for our son, who thought he deserved a dinosaur just like his sister.

The Ornithomimosauria or "bird-mimic dinosaurs" were similar in appearance and probably in mode of life to the ratite birds such as the emu and ostrich. Like them, they were fleet-footed runners, perhaps the fastest dinosaurs that ever lived. Also like the ratites, they were probably opportunistic feeders, living primarily on plants but taking the occasional small animal if the opportunity presented itself. Like modern birds, their ancestry is to be found among the carnivorous theropods. Also like ratites, their jaws were toothless. All widely accepted occurrences of ornithomimosaurs are from the Late Cretaceous of eastern Asia and western North America. That they now have a possible record in the Early Cretaceous of Australia is suggestive that perhaps the group may have originated on the southern continents rather than where they are best known.

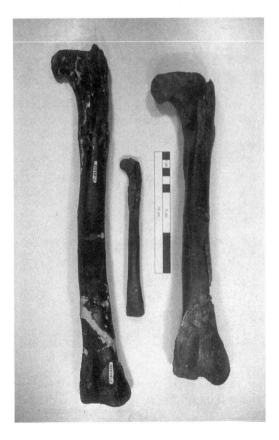

Figure 42. *Left:* adult femur of the ornithomimosaur, or ostrich-like dinosaur *Timimus hermani,* found at Dinosaur Cove. *Middle:* juvenile femur of *Timimus hermani* found within half a meter (1½ feet) of the adult femur. *Right:* adult femur of an ornithomimosaur from the Late Cretaceous, Alberta, Canada.

For the first month, the effort at Slippery Rock was confined to digging out the overburden above the fossiliferous layer in the Western Chamber. It was quite obvious that arduous as this work was, it would be much more so if the formwork and other paraphernalia which would attend construction of the concrete pillar were in place. Therefore, it was decided not to start work on the concrete pillar until at least the bulk of the overburden from the Western Chamber was removed.

For a time, the ancient bogger loaned to the project sped the removal of the overburden along. Then, metal fatigue in the brake bands caused them to fail repeatedly, until half their weight seemed to be that of the welding rods used to repair them. Finally, the more reliable, if more laborious, wheelbarrows were resorted to once again.

After three weeks, only a small amount of overburden remained in the Western Chamber. It was important that the pouring of the concrete pillar alongside the Pillar be completed before the end of the dig in another two months, or else it was likely to be 1995 before the Pillar could itself be excavated. Because it was not certain how long it would take to construct this structure, which was going to be 8 meters (26 feet) long, 1.5 meters (5 feet) wide, and 2.5 meters (8 feet) high, and weigh 60 tonnes (66 tons), Tom wanted to allow sufficient time to complete the job. For $50,000, it could have been done in one day using modern technology.

Fifty thousand dollars was not available, but there was a determined band of volunteers who were. Using a small electric-powered cement mixer that was set up and operated in the First Cross Tunnel, pouring of the cement began on January 30th after two days of digging a foundation for the concrete pillar and constructing formwork.

More than half the job was getting the sand, gravel, and cement to the site. Fortunately fresh water was easy to get because of the high-pressure line that existed to supply the drill steels. The rest was a labor-intensive process consisting of fourteen steps, most of which were carried out with human muscles:

1. Shovel sand and gravel *by hand* into hessian bags or gunnysacks;
2. Load bags of sand, gravel, and cement onto trailer *by hand;*
3. Transport bags to head of track leading to flying fox;
4. Unload bags from trailer and stack *by hand;*
5. Load bags onto miniature trailer behind garden tractor *by hand;*
6. Towing the miniature trailer, drive the garden tractor from the trailer to flying fox;
7. Unload bags from miniature trailer and stack *by hand;*
8. Load bags onto flying fox *by hand;*
9. Send flying fox down to shore platform with load of bags;

10. Unload bags from flying fox *by hand;*
11. Carry bags over shoulders into tunnel and stack *by hand;*
12. Pour contents of bags into cement mixer *by hand;*
13. Pour cement *by hand;*
14. Return hessian bags or gunnysacks to sand and gravel stockpile and start all over again.

Pouring the concrete during the first third of the construction was a straightforward matter of simply tipping out the concrete mixer. As the pillar got higher, it was practical for a time to raise the concrete mixer, but for the last one-third it was necessary to pour the liquid concrete into a bucket and lift it, then pour it into the formwork.

After a month of doing this, the end was in sight. A party was planned in celebration of the completion of the task. A few days before the end, Tom bet every member of the crew a slab (case) of beer that they would not be finished by March 2nd. The bet was eagerly accepted and the crew was so confident of winning that the day before, they took a scheduled break.

The final day began with the crew up and eager to get to work. The feeling was that the job would be easily finished by 4:00 P.M.; then they would have a real party with all that beer. They set to with a will, and in the meantime, Tom organized to get all the beer as well as the makings for a proper celebration. The last thing Tom wanted to have happen was to "win" his bet, because the crew's morale would be devastated. In any case, because Tom does not drink beer, what was he going to do with a dozen slabs of the stuff?

At about 4:00 P.M., Tom went down on site and saw that things were not looking good. The belt kept coming off the cement mixer, and if that machine failed, there was no way the concrete pillar would be completed by the midnight deadline. Tom managed to jerry-build a guide to keep the belt in place and the work continued. Fortunately, the night was warm and the sea was calm. For the first time since the incident in 1984 when the sea had swept away most of the equipment in the dark, Tom allowed work to continue after sundown. It was very pleasant there that night with the languid sea gently lapping at the rocks. About 8:00 P.M., dinner silently arrived on the flying fox, sent down by the magician cooks, Roz and Barry Poole. We all stopped to eat a most pleasant meal sitting on the rocks by the water's edge.

Then it was back to work with sand, gravel, and cement coming down on the flying fox as fast as it could operate. The pouring continued and the hours went by. Eleven P.M. passed, and still there was a significant amount of concrete yet to be poured. The usual banter among the crew died away as they worked ever more intensely to finish the job before midnight. Approaching the highest part of the ceiling required that concrete be jammed in and formwork then be put in place to keep it

Figure 43. Roz Poole.
Photographer: Peter Menzel.

from flowing out. Old bits of cardboard served this purpose as the end was approached—and they held. Finally, at 11:52 P.M., the job was done. An hour was spent tidying up and then the weary crew began the climb out of Dinosaur Cove in the dark.

As they returned to camp, candles were burning in the Recreation Tent and a sumptuous feed had been laid. People made a desultory effort to start a party but were simply too tired to get into a truly festive spirit. Some drank a symbolic bottle of beer and all soon drifted off to bed. About 11:00 the following morning the victory party began and lasted most of the day. Tom cannot imagine ever winning more than by having "lost" that bet.

In the three weeks that remained, the effort returned to completing the excavation of the Western Chamber. During this period, it was fortunate that four vertebrate paleontologists from the Royal Tyrrell Museum of Palaeontology in Canada—Phil Currie, Jane Danis, Darren Tanke, and Kevin Aulenback—could join the field party for a short time. Their visit was made possible by an agreement between their museum and the Museum of Victoria to loan for an indefinite period a skeleton of a hadrosaur, or duck-billed dinosaur, which was complete except for its skull. With many such specimens lying unprepared in their collections, the Canadian museum agreed to make this available

Map of Slippery Rock Site, Dinosaur Cove
as at 27 March 1991

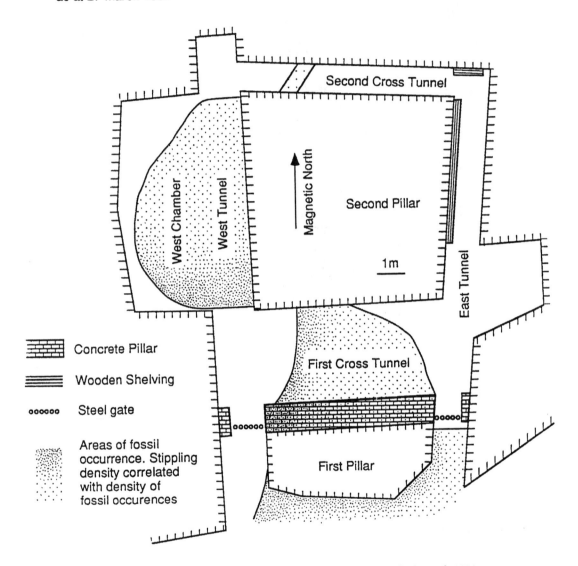

Figure 44. Map of the tunnels at the Slippery Rock site at the end of March 1991.

to the Museum of Victoria as a long-term loan in exchange for providing airfares for four of their staff to visit Australia and participate in the work at Dinosaur Cove.

During their visit, they not only gained firsthand experience of the quite different conditions that prevail in the Australian dinosaur fossil fields but imparted to us and others associated with the project much practical knowledge. Darren, working in the Western Chamber, found the best specimen to be uncovered there, a complete hypsilophodontid vertebra. Jane found a complete fish jaw, one of very few recovered from Dinosaur Cove. After the dig was over, Phil was able to examine the collection that had been prepared from previous digs and, among the many fragments of bone which we could do little with, was able to tentatively suggest that four different groups of small carnivorous dinosaurs may have been present.

In the Western Chamber, the fossiliferous rock was patchy, varying markedly in thickness over a short distance; sometimes it disappeared altogether. Besides the few excellent specimens that the Canadians found, there was little else collected of note and disappointingly little in total. Why the one channel deposit should have been rich in bone and another deposit, which was seemingly identical as far as can be determined, was almost devoid of bone cannot be answered.

Prior to 1991, we had always worked straight through a field season except for one or one-and-a-half days off per week. As an experiment, we built in a one-week break in the middle of the nearly three-month dig. This strategy gave those people who were involved from the beginning to the end a shorter-term goal. It was so successful that in 1993 we built in two such breaks.

6

New Explorations

The Paleobotanists

Robert Spicer is a paleobotanist based at the Open University in Milton Keynes, U.K., and Judith Totman Parrish is a paleoclimatologist at the University of Arizona, Tucson, U.S.A. The two of them have spent more than a decade working together to understand the climatic conditions and flora that existed on the North Slope of Alaska during the Late Cretaceous. That was in the 35 million years between when the rocks where Victoria's dinosaurs are found were deposited and the extinction of dinosaurs as a group. Because of our common interest in polar, terrestrial biotas of the Cretaceous, Judy had come to Melbourne for a conference. Judy invited us to join her and Bob on the North Slope if ever we could get there. We did, in July 1989. This was one of the rare trips that we made together without our children, who this time stayed behind with their grandparents, who are also paleontologists.

The detailed picture of polar biota at the other end of the world that these two people working on the North Slope had been able to pull together impressed us. Other people had worked previously on the Early Cretaceous plants from Victoria, analyzing them from a climatic viewpoint. But when we saw what Judy and Bob had done in Alaska we became convinced that we needed the benefit of the experience they had gained by analyzing the north polar paleofloras from a climatological perspective.

The Committee for Research and Exploration of the National Geographic Society supported a visit by Judy and Bob to Australia toward the end of the 1991 field season in order to make a collection of their own. Once in Australia, they were generously assisted by Jack Douglas, retired from the Victorian Mines Department, who took them to many of the plant fossil sites he knew about. Jack had earlier discovered and collected from them in order to acquire the material that formed the basis for his two-volume work on the Early Cretaceous fossil flora of Victoria, the standard reference for these fossils.[28]

In their subsequent analysis of the shapes and structures of the fossil leaves, Spicer and Parrish concluded that the mean annual temperature during the time when the dinosaurs lived and died in southeastern Australia was probably close to 10°C (50°F), about the same as modern-day London, Chicago, or Denver.[29] This was certainly not the tropics, but indicated warmer conditions than the geochemist's estimates that were based on analysis of oxygen isotope ratios. The reason for the discrepancy is not yet understood, and resolving it is one of the more interesting ongoing research questions of the project.

Figure 45. A fern thriving in modern times during winter in the Australian Alps. Ferns very similar to this one were a major component of the Early Cretaceous flora of polar southeastern Australia.

Interestingly enough, Bob Gregory analyzed a concretion Pat collected on the North Slope for its oxygen isotope ratios and found that the estimated mean annual temperature for that specimen, 10°C (50°F), was identical to what Spicer and Parrish had estimated on the basis of the fossil plants found nearby. Why their results on the Alaskan material should agree while those on the Australian samples are so different could have more than one explanation. It may lie in the fact that the Alaskan Late Cretaceous flora is dominated by angiosperms, or flowering plants, which are all but absent in the Early Cretaceous of Victoria. Most scientific work to date that estimates mean annual paleotemperatures on the basis of leaf shape and structure has centered on angiosperms. Of necessity, Spicer and Parrish pioneered techniques to carry out their investigation of the Victorian floral material, while the Alaskan fossils could be analyzed using established procedures. What Judy and Bob relied on to make their mean annual temperature estimate was the overall shape of the leaves, the proportion of thick to thin cuticle ("skin") of plants, and the ratio of deciduous to evergreen species, together with an informed guess based on the structure of the floral community and how it relates to climate in their experience.

The Rift Valley

A revolution in the thinking of geologists took place in the 1960s and early 1970s. Traces of the idea that the continents have not always been where they are now can be traced back to the sixteenth century, when it was first noted that the east coast of South America and the west coast of Africa "fit" like pieces of a jigsaw. Albert Wegener was the first to pull such vague ideas together and present a synthesis of them, initially publishing his book *Die Entstehung der Kontinente und Ozeane* in 1915; three subsequent editions appeared over the next decade and a half.[30] He noted many similarities between fossils in areas now well separated that could best be explained by the hypothesis that the continents had moved.

Then Wegener made a serious tactical error. If he had stopped with presenting the evidence that the continents had moved, he would have been on solid ground. Had he stopped there, to the geological community as a whole, the cases he had mustered for continental drift would have remained either anomalies to be somehow explained within the broader framework of the conventional geological theory of the day (which held that the positions of the continents had remained as they are today) or as a stimulus to others to work out how the continents did, in fact, move.

However, Wegener went further and proposed a mechanism for continental movement. He was trained as a meteorologist, and his proposed mechanism for continental movement was and is geophysically implausible. A symposium to consider the merits of the idea of continental drift was organized in 1926, and the geophysicists tore that aspect of Wegener's theory apart.[31] As a result, the whole idea of continental drift was discredited in the eyes of the majority of geologists, especially those who dominated the field from the northern hemisphere. The idea would not be revived for another forty years. Ironically, when it was revived, it was the geophysicists who provided the key.[32] During the intervening period, a handful of geologists who were primarily from the southern hemisphere continued to seriously consider the possibility that continental drift had, in fact, taken place. One of the principal advocates was Alexander du Toit, who in 1937 published *Our Wandering Continents*.[33] Based as he was in South Africa, in his day-to-day working life as a geologist du Toit was continually exposed to the geological and paleontological similarities between the southern continents that had so impressed Wegener. So, it is not surprising that he would write such a book rather than a geologist from the northern hemisphere.

In hindsight, it is easy to conclude that the geologists of Wegener's day were heedlessly blind to his ideas for selfish or shortsighted reasons. But this generalization is much too simple. Geologists of the day did give Wegener's ideas serious consideration; otherwise the symposium of 1926 would never have been held. That they did not sort the wheat from the chaff is a reflection of the practical limitations on the amount of time research scientists have to devote to analyzing in depth the various novel hypotheses that are under active consideration in a particular area of knowledge at any one time. Of those, several hypotheses, if true, would cause a revolutionary change in one or more sciences. However, upon closer inspection, most of these hypotheses fail. Any individual scientist has only a limited amount of time to devote to understanding proposed new concepts and testing them critically. Therefore, in sorting through the potentially revolutionary ideas which are about at any one time, many will be rejected by individual scientists, not necessarily because they have carefully thought through the arguments in detail but because when they first learn about them, the theories do not seem plausible enough to devote sufficient time and energy to justify their further pursuit.[34] By linking his idea that continents did move (which is accepted almost universally today) with his idea about how the continents moved (which no one accepts today), Wegener inadvertently cast a pall over the total concept.

The theory that the geophysicists came up with in the 1960s and 1970s for how the continents moved was plate tectonics rather than continental drift. This difference in terminology was not a matter of petty semantics but went straight to the heart of the issue on which Wegener's original concept foundered. Wegener had visualized the continents as analogous to ships formed of lighter rock floating in a sea of denser rock, the rock of the sea floor. As such, the continents moved, in his theory, by sailing through the rock found at the bottom of ocean basins. In Wegener's theory, it was the immense drag caused by this displacement of one rock with another that the geophysicists at the 1926 symposium could not accept (and no one does today).

The theory of plate tectonics, on the other hand, suggests that solid crustal rock is created from molten rock forced upward at mid-ocean ridges. The crust is either pushed out by additional molten rock coming up from below and/or pulled away from the mid-ocean ridges as the

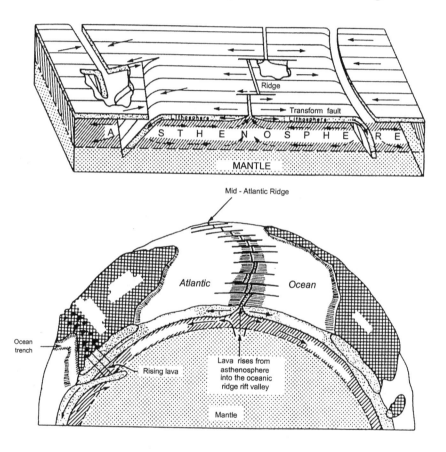

Figure 46. A block diagram and a model of the Earth showing the elements of plate tectonic theory. Ridges are where lava rises from the deeper layers of the Earth to the surface and, once added to the surface, causes expansion of that surface; for example, the Mid-Atlantic Ridge. Trenches are areas where one crustal (or lithospheric) plate dives under another; for example, the trench along the western coast of South America that causes the devastating earthquakes of Chile. Transform faults are areas where crustal plates slide past one another laterally, such as those offsetting the Mid-Atlantic Ridge. It is the relative movement of these crustal plates that brings about the drift of continents. (With the permission of P. Wyllie and L. Sykes; drawings modified from their originals.)

crust at the opposite margin of the same crustal plate is consumed as it sinks into an oceanic trench (subduction zones). Such oceanic trenches are the sites where one gigantic crustal plate dives beneath another. In this way, both the continents and the ocean bottom slowly move away from the mid-ocean ridges at a typical rate of about 10 centimeters (4 inches) per year. Between the mid-ocean ridges and the trenches, the solid crust (the lithosphere) slides over deeper rock that is hot enough to deform the asthenosphere in a plastic manner. The solid and brittle lithosphere forms the upper 100 kilometers (60 miles) of the Earth's crust. The primary evidence for this theory comes from the study of earthquakes and geomagnetism, the provenance of Wegener's harshest critics, the geophysicists.

With the widespread acceptance of the idea of plate tectonics by the early 1970s, many scientists began to devote themselves to working out the movements of the continents during the past. One of their insights was that the boundaries of the continents had changed in the past so that parts of what had once been land attached to what is today northern Australia–New Guinea are now part of Southeast Asia.[35]

The answer to the question of what plate tectonics had to say about Australia when the dinosaurs in Victoria were living animals turned out to be quite intriguing. Australia and Antarctica were closely juxtaposed at the time, and Antarctica was very close to where it is today. Consequently, Australia was far south of its current position. The entire expanse of ocean between the two continents has formed over the past 100 million years as Australia drifted north at a speed comparable to the rate at which fingernails grow.

At the time that our polar dinosaurs were living animals, the process of separation was just getting under way and was far from complete. The rate that the two continents were then pulling apart was much slower than it is today; that is, a few millimeters a year instead of about 10 centimeters annually. Between the main masses of the two continents at this time was a rift valley, much like the Great East African Rift Valley of today. Australia separated from Antarctica in a scissors-like motion, with the split beginning at the southwestern corner of Australia and proceeding eastward. Across the floor of the rift valley great rivers flowed westward toward the eastwardly encroaching sea that would finally reach Victoria 10 million years after the youngest dinosaurs that we know to have lived in Victoria were alive. Smaller streams spilled into these great rivers, and the bones of our polar dinosaurs were entombed in those rivulets. It was in the former beds of some of those small streams that the fossils were found.

The fact that the fossil bones were almost all found in the beds of these former small streams had a profound effect on what we discovered. A few fragmentary bones are all the evidence we have of large dinosaurs. Like the *Allosaurus* astragalus found when the project was just getting started, what specimens we do have of larger dinosaurs are the smallest bits of bone that can be identified as belonging to such animals. Most of what is in our fossil collection consists of the remains of animals that ranged in size between chickens and adult humans. This size bias, which is dictated by the size of the stream channels, is unavoidable, but at least we are able to recognize that the bias is there. When it comes to using these fragments to reconstruct the entire community and trying then to infer the sorts of interactions its various members had with one another, we know there must be other biases which we cannot detect and which will surely distort our analysis. This does not mean we should not attempt such studies; rather, it means that when we do undertake them, we must always be on the lookout for the effects of hidden bias.

The sediments that washed into the rift valley were mostly volcanic debris, the source of which we have no trace except the sediments themselves. The sediment-producing volcanoes were probably to the east, near modern-day Lord Howe Island. It is fortunate that these sediments were first buried and preserved along the Victorian coast long after Australia and Antarctica had separated and were then uplifted in the past 20 million years to form the Otway and Strzelecki Ranges. Only because of this preservation and later exposure do we have accessible rocks where it is possible for us, the dinosaur hunters, to search for fossil bones.

Rocks of the rift valley are known further west in Australia, but all are still buried. We know this from cores of sediment brought up when wells have been drilled in the area. Rocks offshore of East Antarctica, at the other side of the rift valley, have been sampled once. A piston-core sample taken from the ocean bottom not far offshore there in 1979 yielded rock fragments that are no different from the rocks of the Otway and Strzelecki groups. These Antarctic rock fragments also contain fossil pollen of the same species of plants found in those Early Cretaceous rocks exposed along the south coast of Victoria.[36]

In any soft sediment, there are iron and nickel particles bound up in minerals. Because these elements are affected by a magnetic field, they tend to orient with their magnetic axes parallel to the Earth's magnetic field as it is at that moment. As long as the sediment remains soft, the particles will rotate if the Earth's magnetic field changes. But once the sediment turns into hard rock, the orientation of these particles is fixed. Geophysicists have worked out methods to determine the orientation of these particles in hardened sedimentary rocks. In such a way, the former position of the north and south magnetic poles with respect to individual continents can be determined.

The positions of the magnetic poles are not at the rotational poles of the Earth. Rather, today they are displaced from the rotational poles by about 1,340 km (835 miles). Every 3,000 years, the magnetic poles of the Earth migrate around the rotational poles, returning to close to where they were 3,000 years before. Therefore, if the position of the magnetic poles is averaged over several millennia, a reasonably accurate estimate of the position of the rotational poles can be made.

One of the primary tools that has been used to reconstruct the former positions of the continents has been to determine the position of the south and north magnetic poles and, by averaging that, determine the rotational poles relative to each continent. As the continents have moved and rotated, the position of those poles with respect to them has changed. Because the various continents have moved differently with respect to the magnetic poles, if the polar positions through time are plotted for a particular continent, the track of the poles has a unique shape (see Fig. 47, which shows the polar wander path with respect to Australia). An accurate reconstruction of the former relative positions of two continents for a particular time causes their polar wandering curves at that time to overlap.

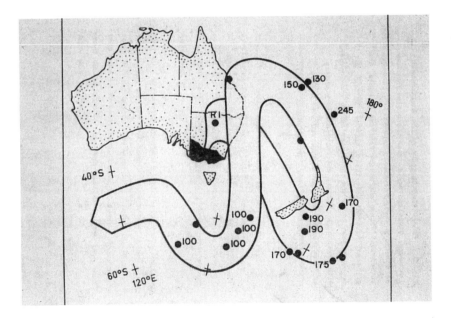

Figure 47. A polar wandering curve for Australia.° The broad band shows the estimated position of the rotational pole of the Earth with respect to Australia during the past 200 million years. The width of the band represents the uncertainty in the determination of the position. Victoria is shaded in black.

°After Schmidt & Embleton (1981) and from Douglas & Williams (1982).

Based on hundreds of paleomagnetic determinations carried out on the rocks in the Otway and Strzelecki Ranges, southeastern Australia appears to have been well inside the Antarctic Circle during the Early Cretaceous when the known dinosaurs in Victoria were alive. This means that for three months of the year the sun never rose on those dinosaurs and their environs.

Return to the Strzeleckis

Michael Cleeland is a resident of Phillip Island, which is not far from Eagle's Nest, where Ferguson collected the first Victorian dinosaur bone. Phillip Island is even closer to San Remo and the Punch Bowl, where Tim Flannery and Tom had earlier found the labyrinthodont amphibian jaw. In 1989, Mike showed up one day at Dinosaur Cove to announce that he was interested in working with us. He spent a few days with us. He rapidly learned how to identify fossil bone in the rock and returned home resolved to search again the outcrops between Inverloch and San Remo. He was thoroughly convinced he would find more fossil bone.

At about the same time, Lesley Kool decided to do the same thing because of the proximity of the area to Melbourne and the fact that after Tim Flannery's efforts there in the late 1970s, no one had been back. Since then, she reasoned, it was quite likely that new material had been exposed by relentless action of the sea washing over the rocks.

That same year, Andrew Constantine, a New Zealander, became a Ph.D. student at Monash University, working with Pat and sedimentologist Ray Cas. His topic was the sedimentology of the Strzelecki Group between San Remo and Inverloch, and he soon acquired a keen eye for fossil bones which seemed to be linked with a latent interest in paleontology. Andrew was already an accomplished field worker and tireless prospector, who liked to carry out his tasks alone. Andrew also has another skill of great value in searching the coastal outcrops: he is a "mountain goat." Many outcrops that seemed to be inaccessible to most mortals were easily within his reach.

Working both separately and together, these three began to turn up fossils regularly. Soon, a number of sites which had never before yielded fossils started to produce material. By the beginning of 1992, with the help of several others who participated in the work from time to time, the three of them had located nine sites which had a realistic potential of yielding enough fossils to rate systematic excavation. We decided to visit all nine localities and, if the prospects seemed good enough on initial inspection, to try digging for one or more days at each in order to evaluate whether fossils were abundant enough to justify further work at any of them.

Several incentives existed to find new sites, particularly in the Strzelecki Group. In the first place, all the known terrestrial vertebrate sites in the Otway Group are about 10 million years younger than those in the Strzelecki Group. Therefore, an extensive fauna from the Strzelecki Group would allow a look at the biota from another time, giving some information about change over those millions of years. Second, different sites might have accumulated fossils in different ways, resulting in different elements of the community being sampled and perhaps more complete specimens being preserved, which would provide additional information about species already known. Third, different sites might sample different environments. Already Andrew Constantine had suggested that labyrinthodont amphibians lived only in environments close to the margins of the rift valley rather than out on the floodplain, perhaps because they preferred faster-flowing water. Two pieces of evidence supported this hypothesis: their fossils had been found near the margins of the ancient rift valley, and the rocks in which they occurred were coarser grained than those which produced the majority of fossil vertebrates, which suggested environments where the streams and rivers had flowed faster than elsewhere.

Of the nine sites examined, three were rejected after an initial visit because there was no indication that the bones were concentrated enough to yield sufficient fossils upon digging in. Of the remaining six, one paid off handsomely, and quite interesting fossils were found at two others. The other three yielded little or nothing after two or three days were spent digging at each.

At Rowells Beach, Mike Cleeland found a left and right mandible of a labyrinthodont amphibian, and Lesley Kool subsequently prepared it. It seemed particularly appropriate that this animal be named *Koolasuchus cleelandi*.[37] In part, the generic name is a pun based on the idea that has been suggested that this group of animals survived in polar Australia long after they had become extinct elsewhere because the environment was too cold for crocodilians. Labyrinthodonts were somewhat like crocodilians in mode of life and body shape and thus, it seems, ecological counterparts. When the climate became somewhat warmer, as evidently had happened by the time the deposits at Dinosaur Cove were laid down, crocodilians moved in, and no remains of labyrinthodonts are known from these younger rocks. Possibly these big amphibians were displaced by the crocodilians. That one of the people involved with this specimen just happened to have the surname Kool was simply too good an opportunity to pass up.

Because both left and right jaws of *Koolasuchus cleelandi* were found together, it seemed quite possible that with more digging, more of its skeleton might be found nearby. That idea, unfortunately, fell into that all-too-familiar category of ideas that are neat, plausible, and *wrong*!

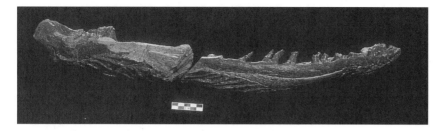

Figure 48. The right jaw of the *Koolasuchus cleelandi* specimen found at Rowells Beach. Unlike the specimen illustrated in Plate 2C, because of the nature of the teeth of this fossil, it is immediately obvious that it is labyrinthodont amphibian.

Digging all around where the jaws were found failed to uncover another scrap. Nearby was a small hillock on which several specimens had been found on the surface. Digging through it to find bones that had not been exposed by erosion again failed to uncover even a single scrap.

A visit to the Punch Bowl, where the first labyrinthodont jaw had been discovered in 1978, resulted in a cluster of bones being found at one end of a large block of rock. As best Tom could remember after fourteen years, this was close to the site where that enigmatic jaw was found. A diligent search of the general area revealed no other fossil bones nearby, so it seems likely that this cluster of bones belongs to one individual.

Breaking off the end of the block where the bones occurred revealed several bone fragments. Rather than risk breaking them up by reducing that piece in size any further, Tom decided to take it out as it was. However, the block of stone was so heavy that it could hardly be moved, much less taken up the steep track to where the vehicles were waiting. Mike Cleeland suggested that getting the block up the track would make an excellent exercise for the local Surf Life Saving unit. Because he knew them, he promised to organize it, and the next week it was done. The "victim" was removed on a stretcher with all the care that would have been given to a human being rescued from that site. The "rescue" of the fossils was definitely in the best of hands!

The next site we explored was The Arch, a natural arch formed out of sandstone along the coast not far from Kilcunda, Victoria. We spent one day there and found only four fossils. One of them absolutely surprised us, however, when it was identified a few months later.

At that time, if we had been asked what groups of dinosaurs would never be found in Australia, the neoceratopsians would have been at the top of the list. Previously known only from much younger rocks and primarily from eastern Asia and western North America, these frill-necked and often horned dinosaurs seemed to be the most geographically restricted of all these reptiles.

The protoceratopsians are small members of the Neoceratopsia or "new-horned dinosaurs," larger members of which include the well-known *Triceratops,* a member of the exclusively North American Ceratopsidae. These have well-developed horns, the feature which is the basis for the name of the entire group. In contrast, *Protoceratops* itself has only a weak nasal horn, and many other protoceratopsians lack any sign of a horn at all. The neoceratopsians were the last major group of dinosaurs to appear in the fossil record. With the exception of a few presently unlocatable specimens from the latest Cretaceous of South America, they are known primarily from the Late Cretaceous of western North America and Asia, where they are very diverse, particularly the horned Ceratopsidae.

A few months after we found the fossil at The Arch, we visited the Royal Tyrrell Museum of Palaeontology in Alberta, Canada. There we tried unsuccessfully for several days to convince ourselves that this bone was an ulna, or lower arm bone, of a dinosaur belonging to one of the groups that includes forms such as *Tyrannosaurus* or *Allosaurus*—they have small forearms and this fossil had short, stocky proportions for an ulna. However, when the specimens of these carnivores were in front of us, we realized that while they are small, their ulnas are proportionally elongate rods rather than short and stocky. In the meantime, purely by chance, Darren Tanke, who had been to Dinosaur Cove the year before, had laid out a skeleton of *Centrosaurus,* a near relative of *Triceratops,* on a series of nearby cabinets. The physical proximity of this skeleton almost forced us to look at it, even though such dinosaurs were far from our minds at the time. Finally, it dawned on us that this specimen of *Centrosaurus* had an ulna surprisingly like the one from The Arch, except that it was about four and a half times as long. Becoming rather excited by the seemingly unlikely possibility that the specimen was a neoceratopsian, we asked if there were any of the smaller forms more like *Protoceratops* available. There were none in their collections, but nearby in Drumheller there just happened to be a skeleton of *Leptoceratops gracilis* on loan from the National Museum of Canada. That specimen had been collected only a few tens of kilometers north of Drumheller. When we placed the ulna from The Arch next to this specimen we saw at a glance that the two fossils were strikingly similar. Had they been found in the same hole in the ground, they would probably be regarded as extreme variants of a single species or as closely related species, for the Victorian fossil is slightly shorter and deeper but otherwise virtually identical. Yet they were separated by 13,000 kilometers (8,000 miles) and 50,000,000 years (1,300,000,000 fortnights!).[38]

Unfortunately, to dig a significant quantity of rock from the base of

The Arch would not only speed up the destruction of an unusual geological feature, it might also be dangerous to the person doing it. One can only hope that at another site, more definitive evidence for the presence of Australian neoceratopsians will eventually be found. A hint that there is more to be found comes in the form of another ulna found at Dinosaur Cove, which shares the unusual proportions characteristic of neoceratopsians but is only about half the size of the one from The Arch.

When one or two specimens of this nature point to a major revision of the ideas concerning the course of evolution of a large group, it is prudent to consider other explanations. The only alternative that has occurred to us seems rather ad hoc, but nature does not always follow the rule of Occam's razor, which states that the simplest explanation is most likely (but not necessarily) the true one. The alternative that has come to mind is the existence of a group of reptiles, perhaps dinosaurs, with an ulna that closely resembles the otherwise unique pattern of the neoceratopsians but is in other respects quite different. Hypothesizing the existence of a whole new group in order to explain two fossils does seem an extreme alternative to simply extending the range in time and space of a known one. So, the case rests until more material is found. Since that first specimen of a possible Australian protoceratopsian was recognized, protoceratopsians about as old have been reported from Asia[39] and North America.[40] At present, although the Australian specimen is still a geographic surprise, it is no longer significantly older than the oldest records of that dinosaur group elsewhere.

Between Eagle's Nest and the town of Inverloch is a stretch of coast called Flat Rocks. This was the last of the potential sites that we surveyed. People searching for fossils had walked over this area more than once, for it is one of the more accessible stretches of coast where they might be found. Little or nothing turned up until one day after a major storm had swept away the 10–20 centimeters (4–8 inches) of sand that normally buried much of the shore platform. On that propitious day, Lesley Kool, Mike Cleeland, and some associates were walking over the area. In a band of rock a few meters wide that lay in an arc across the shore platform, they found twenty to thirty exposed fossil bone fragments. The next tide buried this cornucopia of fossils, but its location was indelibly recorded in their minds. It was now a simple matter to return to the locality and, in a few minutes, shovel away a bit of sand to expose part of the fossiliferous layer once again.

The arc across the shore platform marked the course taken by a small stream about 115 million years ago. That it just happens to be at the level cut by the sea and exposed so exquisitely is good fortune. If the sea level were 1 meter (1 yard) lower, the situation would be identical to that at Slippery Rock. Were it 1 meter higher, the ancient channel would still be underwater and buried in rock.

The first day of experimental digging there was so successful that twelve more days were devoted to working that one site. On average, the daily yield was forty to sixty fossil bone fragments. That was as good as the most productive days at Dinosaur Cove. Equally pleasing was the fact that while nearly a decade was required to dig out 20 tonnes (22 tons) of fossiliferous rock at Dinosaur Cove (because another 600 tonnes of overburden had to be removed from tunnels to reach much of it), here at Flat Rocks there were at least 40 tonnes of fossiliferous rock with no overburden whatsoever!

Instead of descending 90 meters (300 feet) down a track that can be notoriously slippery in spots (as at Dinosaur Cove), the descent to Flat Rocks is about 8 meters (25 feet) down well-maintained wooden steps and across a small wooden bridge. Heaven!

At Flat Rocks, in ten minutes all the equipment can be transported from our vehicles to the beach and digging can begin. This was a most welcome change from Dinosaur Cove, where days were spent setting up equipment at the start of each field season before collecting could even start. Also at Dinosaur Cove one confronted an equally arduous take-out at the end of the dig, when many volunteers seem to find it necessary to depart unexpectedly early, leaving the leader and a few stalwarts the task of cleaning up everything on their own.

Had the Flat Rocks locality been known in 1984, there is no way on earth the Friends of the National Museum of Victoria could have persuaded Tom to work at Dinosaur Cove. With this much easier site to work, it would have been madness to have even contemplated going to all the trouble to organize the underground operations at Dinosaur Cove. Perhaps it is just as well that things happened the way they did, for the two assemblages complement one another, given their age difference of about 10 million years.

Measuring the Paleolatitude

Time and again, in both technical and popular publications as well as in media reports, we have talked about the polar dinosaurs of southeastern Australia. The basis for our assertion that these animals did indeed live and die within the Antarctic Circle of the day is based on a consensus of the work of others. The several reputable estimates published in the last twenty years for the position of southeastern Australia between 100 and 125 million years ago differ from one another quite significantly, although all agree that the area was at a high latitude. Different data have been considered in arriving at these various results.

Previously, only a few measurements throwing light on this problem had been taken from the rocks where the dinosaurs were found. Michael Whitelaw had been at Dinosaur Cove for the first excavation in 1984 and organized and ran the one in 1986. He had been Pat's undergraduate student and then went to work with Bruce McFadden and Neil Opdyke at the University of Florida. Subsequently he received a Ph.D. there for work on paleomagnetism. As we have seen, by means of studies of this nature, the former latitude of a region can be determined.

Mike carried out an extensive sampling program of the rocks immediately adjacent to where the dinosaurs occurred. He estimated that when the rocks at Dinosaur Cove were laid down they were practically at the Antarctic Circle (66° south), while those 10 million years older at the Flat Rocks site were deposited when that area was 10° inside the Antarctic Circle (76° south).

It is unlikely that Australia drifted more than 1,000 kilometers (600 miles) away from the South Pole during that 10-million-year period as Mike's data suggests. The difference in the two figures is probably partly due to movement of the continent but also partly to measurement error. The results do, however, clearly support the idea that the dinosaurs of southeastern Australia were indeed polar animals. As such, they had to contend each year with months of darkness for twenty-four hours per day except, of course, for moonlight and the ghostly glow of the aurora australis.

7

Restoring Life of the Past

A year before Tom began working at the Museum of Victoria, a 12-year-old girl, Kerry Hine, was playing in her family's kaolinite quarry near Bacchus Marsh, a town about 40 km (25 miles) west of Melbourne. A fossil bone caught her alert eyes, and she took it to her science teacher at the local school. He contacted the National Museum of Victoria. Some staff members came out to the site and recognized that here was an important fossil locality.

Digging in the bright white clay in summer was a daunting task. But the crew continued on and collected about a dozen partial skulls (one nearly complete) of the large Pleistocene (between 10,000 and 1,800,000 years old) marsupial *Diprotodon*. *Diprotodon australis* is the largest marsupial that ever lived, but the Bacchus Marsh specimens belong to a smaller variety, perhaps a different species. Although not the largest individuals of *Diprotodon*, the Bacchus Marsh specimens were exquisitely preserved. At other places where *Diprotodon* skulls have been found, they have typically been crushed; the Bacchus Marsh skulls, although often damaged and incomplete, were not crushed.

The skulls of *Diprotodon* are generally crushed because many of the bones of the skull are remarkably thin. In parts of the skull such as the temporal region, the bone is only a few millimeters (⅛ inch) thick. In a comparable-sized placental ungulate such as a hippopotamus, the bone in the same region is typically 20 mm (⅘ inch) or more thick. This incredibly light skull may have been an adaptation that helped *Diprotodon* economize on mineral resources in a continent where soils are generally not rich.[41] It also lightened up the large skull.

When Tom began work at the Museum of Victoria, studying these Pleistocene fossils was not something that he wished to do. His heart was set on finding the older record of birds and mammals in Australia, not investing the time and effort to study these big, relatively recent specimens, because it was obvious that much would have to be done before they could be scientifically analyzed. First, the surrounding rock would have to be removed and the bones hardened. Then an artist would have to be hired to do a meticulous job of illustrating the material. As one year led into another and no one found the long-sought early Cenozoic and Mesozoic birds and mammals, it was suggested to Tom that he could probably find support to do the preparation and art work if he sought it from the Australian Research Grants Committee. So in the late 1970s Tom obtained some grant money, and Tim Flannery and Pat prepared the specimens.

At this time, we lived in Emerald, Victoria, a country town 59 km (37 miles) from the center of Melbourne and the National Museum of Victoria. That was a good hour-and-a-half commute away. Pat then had a part-time job at Monash University but did not go into the city each day because the first of our children, Leaellyn, was still a toddler. Tim also lived in Emerald, so they did the preparation in a renovated shed on our property. This was a godsend to Pat because she was able to eliminate the commute hours, and with the rug we put down in the well-ventilated shed, our daughter could play with her toys while her mother and Tim prepared fossils.

Once the *Diprotodon* specimens were free of the rock, a qualified artist had to be hired to prepare images for scientific study and eventual publication. But how to find an artist both qualified and inclined to make photorealistic drawings with meticulous accuracy? We could not use a photographer because what was needed was someone who could synthesize in his/her mind information from a dozen different broken specimens, then draw a single, accurate composite. What was fortunate about all these specimens of *Diprotodon* was that they were extremely uniform. It was as if a herd of one sex all about the same age had died together.

Today, when a population of kangaroos has suffered through a prolonged drought, the last individuals to survive are those of prime breeding age. In the population of *Diprotodon* from Bacchus Marsh, there were no old individuals or juveniles. In every one of them, the last molar is only partially erupted, so they were all young adults and all remarkably close to the same size. In other places where *Diprotodon* fossils are found, typically there are two different skull types: a larger form with a more robust appearance, presumably the males; and a smaller form that is more gracile, the females. The Bacchus Marsh forms are reminiscent of the more gracile forms found elsewhere.

Just as the skulls were almost completely freed from the rock, there was an exhibition of the works of wildlife artists of the Melbourne area. We attended this exhibition for the express purpose of finding an artist whose style was best suited for illustrating the specimens of *Diprotodon*. As we walked through the exhibition, we noticed the work of one artist and knew immediately that here was the person for the job.

What we saw were paintings of birds so lifelike we could imagine that the living animal was about to jump out of the canvas. That photographic quality was just the approach that would be needed to illustrate the *Diprotodon* fossils.

But would the artist take on the illustration of dead, inanimate bones instead of living animals? The artist, Peter Trusler, was in the room, and after the first introductions, it was clear that here was a person devoted to art who was not only capable of producing the kind of work that was needed but also eager to take on the job. A portrait and wildlife artist, Peter also has a university background in zoology. That technical background was to serve him in good stead repeatedly in the years that followed, because it gave him a superb theoretical understanding of anatomy that added to the practical experience he had gained from countless hours of observing living animals.

The commission to paint the skulls of *Diprotodon* was spread over three years and resulted in the production of a total of fifteen lifelike illustrations. These were so meticulously done that it is possible to take quite accurate measurements from the illustrations. This was made possible in part by a stroke of good luck, the remarkable similarity in size of all the specimens. Peter was able to combine the several specimens in his mind and come up with a scientifically accurate composite illustration that was more complete than any one individual.

From time to time over the next decade, illustrations of other specimens followed. Then in 1989, we began to write the book *Wildlife of Gondwana*,[42] a recounting, with an emphasis on Australia, of the 450-million-year history of vertebrates on the Gondwana continents. The approach we took was to illustrate as many fossils as possible with the highest-quality color photographs and to weave the text around them. But to make some of the fossils come to life, we asked Peter to produce three paintings. One was that of *Leaellynasaura* from Dinosaur Cove.

With some artists, you can supply a list of items you wish to show in a picture and that is the end of the matter until an advanced sketch is produced. However, Peter does not operate in that fashion. He first spent a considerable amount of time just thinking about the final images. We talked a lot, he read our papers, looked at the bones, and borrowed numerous books from our library about other dinosaurs. We discussed what dinosaurs should be portrayed and how they should be posed in their habitat. As he did this, he consulted frequently with us, both about our ideas of how the dinosaurs were to be portrayed and about technical details of the dinosaurs and their surroundings.

After several months, Peter hit upon the idea of portraying the type specimen, or holotype, of *Leaellynasaura amicagraphica* as it would have appeared shortly after death. This solved a number of problems, the first being the stance. Because Peter had never before reconstructed a dinosaur as a living animal, a death pose was easier for him than a living one. It also allowed a very specific statement to be made about the exact habitat at the Slippery Rock site where this specimen was found. We had reconstructed that habitat in detail as the dinosaurs were excavated. When *L. amicagraphica* was a living animal, the site where its bones were found was at the outer edge of an oxbow or billabong pool that was 5 meters (16 feet) across. A number of fragments of other dinosaurs, lungfish, and plant debris were associated with the site. The information about the physical nature of the site was supplied in detail by Pat's graduate student, Andrew Constantine. Although his main work was in the Strzelecki Ranges, he came to Dinosaur Cove on numerous occasions, producing detailed maps and paleoenvironmental reconstructions as we dug. His input was invaluable.

Peter took in all this information and started by sketching *Leaellynasaura amicagraphica* as a corpse. He did this by first reconstructing the skeleton, then reconstructing the muscle masses that would have linked the various bones together. He studied both the muscle scars, or points of attachment of the muscles to the bones of the fossil, and the muscle placement in the crocodilians to make the reconstruction. He also studied dead animals that he chanced to encounter to become familiar at first hand with the process of rigor mortis. With a detailed sketch in hand, he then made a clay model in order to better appreciate the three-dimensional aspects of the corpse. It is a pity that this beautiful model was never fired; it dried out before that could be done and ended up as a pile of clay chips.

In order to reinforce the idea that this animal had lived at polar latitudes, we decided to portray it as if it had died in the autumn, when ice had started to form on the surface of the billabong. While working on preliminary sketches of the animal itself, Peter also studied the effects of light on thin sheets of ice and frozen mud. To understand what plant debris might have looked like sticking through the ice, he froze modern *Araucaria* (or monkey puzzle) fronds, a relative of a plant that thrived millions of years ago at Dinosaur Cove. He obtained modern leaves of the maidenhair tree, or ginkgo, and cut wedge-shaped pieces out of the leaves so that they were closer in shape to their near relative in the Cretaceous, *Ginkgoites*. Then he put the ginkgo leaf pieces on top of a bed of mud and froze the whole thing. Gael, Peter's wife, had to do with a bit less freezer space and store her frozen peas next to a mud-encrusted bunch of leaves!

To show the characteristic light of the polar sky and yet keep the holotype of *Leaellynasaura amicagraphica* large enough to see, Peter could not show the sky directly, so he did it as a reflection in the water. Similarly, had he shown the width of the billabong directly, the image of *L. amicagraphica* would have been smaller than Peter desired, so he illustrated a reflection of the trees on the opposite bank in the water to give a subtle impression of the dimensions of this former water body.

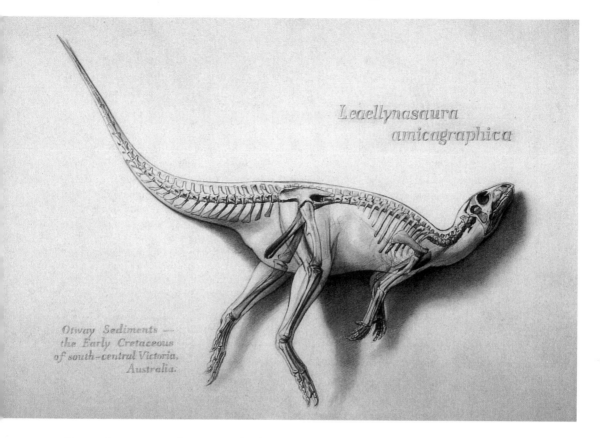

Leaellynasaura amicagraphica

Otway Sediments — the Early Cretaceous of south-central Victoria, Australia.

Figure 49. An intermediate step in the drawing of *Leaellynasaura amicagraphica* that appears in Plate 13.

The actual painting of the final oil took about one-third of the six months Peter spent on the project. As he progressed with it, he continued to quiz us about several aspects of the scene and modified it from time to time as he went.

We were most fortunate at this point that Bill Templeman (publishing director of Reed Books) was the publisher we were dealing with. He provided the funds as part of the book royalties that allowed Peter to complete this exquisite work.

In 1993, Australia Post wanted to have a stamp issue centered on Australian dinosaurs. This was because of all the interest in dinosaurs created at the time by the release of the movie *Jurassic Park*. Australia Post first approached Pat for ideas and she immediately suggested that we work with Peter to produce something unique and new. She also offered to produce an exhibition of the dinosaurs Peter would paint for the headquarters of Australia Post in Melbourne. Our association with Australia Post was not limited to the production of the stamps. As director of the Monash Science Centre, Pat also co-wrote with her staff an education kit featuring the Stamp Gang, which represents Australia Post to Australian kids. And at the end Australia Post was the first group to donate monetary support to help with The Great Russian Dinosaur Exhibition that also opened along with the Australian debut of *Jurassic Park* in Melbourne in August 1993. This was an association that benefited everyone concerned.

Figure 50. The block of Australian dinosaur stamps issued in October 1993. Except for the presence of *Muttaburrasaurus* (the largest dinosaur pictured), this is a scene from the Victorian rift valley as it might have appeared about 110 million years ago. From left to right, the individual stamps are centered on the dinosaurs *Leaellynasaura* (a hypsilophodontid), an allosaurid, *Muttaburrasaurus* (an iguanodontoid), *Minmi* (an ankylosaur), and *Timimus* (an ornithomimosaur). In the background between *Muttaburrasaurus* and *Timimus* can be seen a second hypsilophodontid, *Atlascopcosaurus*. Overhead is the pterosaur *Ornithocheirus*. *Artist: Peter Trusler.* Courtesy of Australia Post.

The creative artistic documentary process was much the same as before. By now Peter was willing to try to paint dinosaurs as living animals, not just corpses. Australia Post gave him a grid pattern which would be the places where stamps would be cut out of a scene Peter would create. Peter's first task was to come up with a plausible layout of the dinosaurs in such a fashion that there would be one dinosaur on each stamp. We worked with him and together came up with the idea of illustrating the rift valley formed between Australia and Antarctica as the two continents began to separate. As that was the place where all the dinosaurs from southeastern Australia were found, it seemed most appropriate. Because Australia Post wanted to have an Australian, not a solely Victorian, dinosaur stamp issue, *Muttaburrasaurus* from Queensland was included among the six stamps. No trace of a dinosaur similar to *Muttaburrasaurus* has yet been found in Victoria. The four other dinosaurs and one pterosaur portrayed were much better known else-

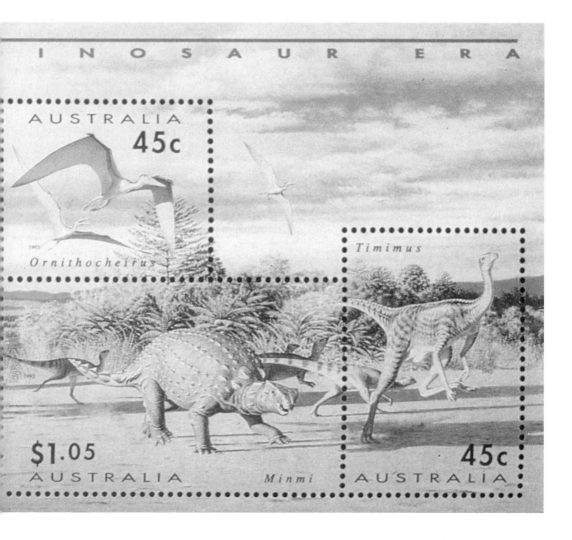

where than in Victoria, but there was a record of each in southeastern Australia. Many of the finest works of the Italian Renaissance were painted with far more strictures on the subject matter than this, so we all felt quite happy with the result, despite having to include *Muttaburrasaurus.*

8

The First Last Excavation of Dinosaur Cove

Just before going to the Strzeleckis to test prospective sites in 1992, we made a brief visit to Dinosaur Cove. The purpose of the visit was to blast off some of the rock that would otherwise have cantilevered out more than 2 meters (6 feet) beyond the concrete pillar once the Pillar was removed. By doing this a year before the planned start of operations in 1993, we gave the sea a winter to pound at the newly exposed surface and bring down rocks loosened by the blast.

After the blasting, Tom was repeatedly scratched by the vegetation along the slippery, rain-sodden access track as he hauled gear out of Dinosaur Cove. A miniscule but deep cut he did not notice at the time among all the superficial ones he was receiving became septic and he developed a severe case of blood poisoning within sixteen hours. As a result, Tom spent four days enjoying the hospitality of the Colac Hospital. Had antibiotics not existed, he would have been extremely lucky to have survived and only lost a finger.

With Tom laid up in hospital, Pat was left to pick up the final bits and pieces of the camp with two kids in tow. Fortunately, as if by magic, David Denney and his son Greg happened to come by and helped her pack up, store all the tools, and stow the last tent. Many times that day she thought quietly to herself of the classic Australian bush story *The Drover's Wife* by Henry Lawson and felt herself lucky that she had help nearby and could soon get away from the mildewed shack we had stayed in at the camp on that visit and return home to sleep in a warm, dry place.[43]

In our initial plan for returning a year later to Dinosaur Cove we envisioned digging out all the remaining fossiliferous rock and permanently closing down the site, at least as far as we were concerned. Who can say what some person, perhaps whose grandparents are not yet born, will want to do there? But for us at least, when the work started in 1993, the goal of excavating all the fossiliferous rock that it was practical to obtain from there seemed an attainable objective. We were wrong.

The Excavation of 1993

With all blasting at Dinosaur Cove, it was rarely as simple to carry out a plan as it first appeared to be. The Pillar was no exception. With one side inaccessible because of the concrete pillar next to it, the fractured basket case with numerous split sets driven through it and steel mesh attached to it proved more durable and irascible than anyone thought possible. Pat O'Neill, now no longer with the Victorian government, carried out the job that he had for six years felt almost surely would have to be done sooner or later. Bit by bit, the Pillar was slowly reduced until after eighteen days it was gone. Because of the danger of rocks falling on people who were working where the Pillar had been, Pat O'Neill had a protective wooden portico built. The 10 square meters (100 square feet) of fossiliferous rock exposed in this way was surely the hardest-won bit of ground in Dinosaur Cove, for not only did the overburden above it have to be removed but the concrete pillar had to be poured alongside it first in order to safely dig that area.

Only when Pat O'Neill was satisfied that the area where the Pillar had been was made as safe as possible were the eager fossil hunters finally allowed to go after the rich accumulation of fossils we were expecting. Some fossils indeed were found, but the yield was disappointingly low. Three and a half years later a specimen collected there was prepared; the rock around it was removed to fully expose the fossil bone. When collected in the field, it had been labeled "Turtle? Humerus?" Its identification as a humerus, the upper bone of the forearm, turned out to be correct, but it was not a turtle. Rather, it turned out to be a monotreme. This discovery made all the effort to build the concrete pillar and remove the Pillar worthwhile, for it was the first mammal recovered from the Victorian Cretaceous. *At last*, we had finally obtained the kind of fossil that was the initial objective of this project we began so many years ago.

With the attainment of this primary goal, however, we were presented with a somewhat unusual problem. Several years previously, Helen Wilson, a stalwart volunteer, had asked Tom what he would give to the person who found the first mammal specimen at Dinosaur Cove. At the time it had seemed an extremely unlikely event, and so he rashly promised the type of inducement that would spur Helen on: one cubic meter of chocolate. Unfortunately, when the monotreme humerus was found and was still mostly enclosed in the rock, it was such a nondescript specimen that whoever discovered it did not bother to record their name on the package in which it was wrapped. Thus, when the one tonne of chocolate was finally obtained, it was divided equally between all the people who had worked at Dinosaur Cove and could make it to the Cadbury chocolate factory where it was given out.

Obtaining the chocolate to honor the promise was a long and drawn-out process. As with Bill Loads and Atlas Copco, it was a case of making the right connection completely by chance. A longtime volunteer, Cindy Hann, is a dedicated science teacher when she is not cracking rocks to look for fossils on one of the digs. One of her students was the son of Frank Miller, who was in charge of the Cadbury operations in Ringwood, Victoria. Fortunately, that monotreme humerus is a dark brown in color, so a name for it alluding to the role that chocolate played in its discovery will not be hard to coin.

In 1957, P. J. Darlington published a book entitled *Zoogeography: The Geographical Distribution of Animals.* In it, he remarked that when one compares the living monotremes, the platypus and echidnas, they are about as different from one another in their mode of life as are the placental otters and porcupines.

Although otters and porcupines are both placental mammals, they are not particularly close relatives. They are, instead, widely separated twigs on the tree of placental evolution, a tree that also includes such different animals as deer, whales, bat, shrews, and monkeys. Darlington suggested that there was probably a similarly diverse evolutionary radiation of monotremes and that all that is left of it are these two groups who are as far apart on the monotreme evolutionary tree as otters and porcupines on the placental tree.

Darlington's idea came to mind as we examined the humerus because, while clearly that of a monotreme, it differs significantly in one feature. At the base of the humerus in many placentals and marsupials there is a pit, the olecranon fossa, that allows space for a projection on the ulna, the larger bone of the forearm. The resulting mechanical arrangement enables the forelimb to be extended straight out while at the same time there is a sturdy articulation at the elbow. None of the living monotremes or any of the other known fossil monotremes have such a pit. The living monotremes have a sprawling posture. The presence of a pit in the humerus from the Pillar suggests that perhaps this animal had a forelimb that was held more upright, as it is in the majority of placental and marsupial mammals. If so, it would have been a monotreme that was, indeed, quite different from the platypus and echidnas. It may well be that this is one of the missing monotremes that Darlington had in mind.

For the balance of the 1993 season, while rock was being taken up where the Pillar had been, further underground work at Slippery Rock continued. This excavation was just to make sure that there was nothing there which had been overlooked before the tunnels were sealed. An area on the north side of the First Cross Tunnel and another one on

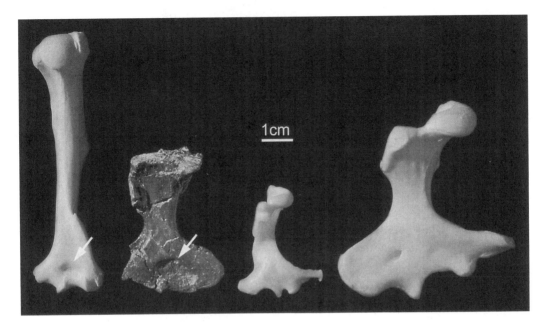

Figure 51. Four right humeri seen in posterior view. The dark humerus second from the left is that of the Early Cretaceous monotreme from Dinosaur Cove. To its left is one of an extant marsupial possum, *Trichosurus vulpecula.* Second from the right is that of the platypus, *Ornithorhynchus anatinus,* an extant monotreme. Farthest to the right is that of the echidna, *Tachyglossus aculeatus,* a representative of the other extant family of monotremes. The location of the pit, or olecranon fossa, suggestive of upright posture in the Dinosaur Cove humerus is indicated by a white arrow in both it and the marsupial humerus. The Dinosaur Cove humerus tells us that monotremes quite different than the platypus and echidna did exist and perhaps had a more upright posture. Its presence at Dinosaur Cove also shows that at least one monotreme was capable of surviving in a polar environment.

the south side of the Western Chamber were exposed by blasting off the overburden. As we dug into the floor at those sites, a few fragments of bone were found in the First Cross Tunnel, but they were a mere shadow of what had been recovered nearby in 1987 and 1989. Not a scrap was found in the newly exposed area of the Western Chamber.

With this done, preparations were made to seal the entrances to the tunnels at the Slippery Rock site. Before doing so, as a guide to anyone who some time in the future might enter those tunnels, a glass jar containing printed information about the site plus casts of some of the dinosaur bones found there was left behind near the junction of the East Tunnel and the Second Cross Tunnel.

Ray Blanford and his helpers then poured concrete to block the entrances of both tunnels. By himself, Ray lifted a black granite monolith weighing 100 kgm (220 lbs.) and placed it in the concrete plug filling the entrance of the East Tunnel. On that stone was inscribed the following words:

SIGNIFICANT FOSSILS WERE

DISCOVERED AT THIS LOCALITY,

DINOSAUR COVE, IN 1980.

FIELD PARTIES COMPOSED

PRINCIPALLY OF VOLUNTEERS

FROM MONASH UNIVERSITY,

THE MUSEUM OF VICTORIA

AND EARTHWATCH COLLECTED

DINOSAURS AND OTHER

VERTEBRATE FOSSILS FROM

THREE SITES WITHIN THIS COVE,

1984–1993. MAJOR SUPPORT CAME

FROM THE NATIONAL GEOGRAPHIC

SOCIETY, ATLAS COPCO, I.C.I.,

THE DEPARTMENT OF CONSERVATION

& NATURAL RESOURCES, AND

THE AUSTRALIAN RESEARCH

GRANTS COMMITTEE

Donated by A. Giannarelli & Sons, Fitzroy

Figure 52. The black granite monolith on which was inscribed a record of what had been done in Dinosaur Cove.

With the sealing of those tunnels, Tom turned, took a few steps, and, looking out to sea, suddenly felt as if a tremendous weight, which he had not even been aware of and which had been there for years, was lifted from his shoulders. Neither of us had ever been happy underground, and it was good to know that no one had ever been permanently injured, much less killed, in the search for dinosaurs in the tunnels. Every time someone drives a car to do something on our behalf, there is a risk of an accident. However, because people commonly drive in their day-to-day lives, that risk was always easier to live with than were the risks of the same people going underground on our behalf in search of fossils. These feelings are there because the devotion of volunteers engenders a feeling toward them as if they were part of your own family. And, like parents, we worry.

The Premature Extinction of Dinosaur Cove

At the beginning of the 1993 dig, the expectation was that the fossiliferous layer at Dinosaur Cove East would pinch out and be exhausted by the time work was finished in early April. As we planned the operations for that year, one of our worries was that both it and the Slippery Rock site would be exhausted well before the planned termination in early April. This would have been a source of embarrassment because the Museum of Victoria planned to have an "Extinction of Dinosaur Cove" party to mark the end of this major project. Thinking this just might happen, the date for that event was brought forward one month. It need not have been.

Rain beat down steadily in torrents all through the night prior to the "Extinction" party. We resigned ourselves to a rather soggy affair. A helicopter was hired to ferry people in and out of the cove that day. As the helicopter lifted off from Warrnambool, the clouds broke up. Intermittent heavy clouds and occasional rain squalls, just enough to prevent first-time visitors from getting the idea that the conditions at Dinosaur Cove were idyllic, characterized the otherwise rather pleasant day. The participants included many old hands who returned to the site to see it once again before it would close down forever, as we thought then. Other participants were old media friends, sponsors, and local residents who had assisted more indirectly, but in very important ways. Many of the latter two groups were seeing Dinosaur Cove for the first time.

The helicopter proved a boon, because it transported many people into and out of Dinosaur Cove who otherwise could not have entered it. Most of the digging crews who had labored there for years had never previously flown in and out of the site, and they immensely enjoyed the opportunity to do so. Through most of the day, the chopper continuously shuttled people back and forth. As it dropped off its last few loads of passengers, the rising tide was covering the landing platform. Finally, the pilot decreed, "Enough!" and thereafter he took people on a fly-by, nearly skimming the waves in the process.

While the partying and helicopter rides were under way, many of the old hands made one last assault with picks and hammers at Dinosaur Cove West. Their nostalgic activity was rewarded with numerous isolated bones recovered from the last of the workable exposures there.

A series of awards was presented at that party to people who had made outstanding contributions to the work at Dinosaur Cove over the past decade. John Herman received the title of "The Mayor of Dinoville"; David Denney, the "Landlord of Dinoville"; and Roz and Barry Poole, the "Chefs of Dinoville." Michelle Colwell was awarded for being the longest-serving crew member (she had been on every dig since the beginning), and Arja Byrne was awarded for delivering FAX messages to the camp on horseback.

One person who was not there to receive his award was John Angel, principal of the Lavers Hill Consolidated School. His award was for "Providing Showers for the Denizens of Dinoville." Over the years, the showers at that school had been used by crew members about 10,000 times. Without them, it would have been extremely difficult to find volunteers so dedicated that they would have been willing to get covered in dirt, mud, and oil every day without a proper wash.

As the last visitors left toward the end of the day and the helicopter flew back to Warrnambool, the solid cloud mass returned and a steady rain beat down again that night. The Fates were definitely on our side that day.

1.2 Tonnes!

Popularizing the work we do takes other forms in addition to the production of radio and television documentaries. One of these alternatives is popular articles for magazines.

Generally there is a fee involved. In one case, we charged $1. This was so that Peter Trusler could be paid $2,999 to produce a painting for the article that was published in the *QANTAS Airways Magazine.* As these magazines are distributed free on the airplanes, Tom asked if we could have some of the copies remaining when the two-month period in which the magazine would be made available to the QANTAS (Queensland and Northern Territory Aerial Service) passengers was over. The reply over the telephone was a typically Aussie "No worries." A few months went by and nothing happened. Then one day, Tom received another telephone call from QANTAS. This time it was from the freight depot at Melbourne airport asking when he was going to pick up his two pallets of freight. "Two pallets? How much is that?" Tom asked. "Oh, about 1.2 tonnes," came the nonchalant answer.

Because what was on those pallets were 12,000 new copies of the magazine, not the few hundred shopworn ones that Tom had hoped for, we had to hire a large truck to pick them up. Ever since they have proven useful as giveaways to children and adults who want to know something about the dinosaurs that lived not far from their homes. This was our first encounter with the generosity of QANTAS. Later cooperative efforts were even more generous, which eventually led us to honor the Queensland and Northern Territory Aerial Service in a special, scientific way.

The Excavation of 1993 (continued)

Initially under the direction of Nick van Klaveren, the crew working Dinosaur Cove East forged ahead, uncovering a small but steady stream of specimens. As they worked north, the fossiliferous unit did not thin out as Tom had thought it would on the basis of his interpretation of cores taken by Rob Anderson in 1991. Instead, it quintupled in thickness to more than 1 meter (1 yard). Never in Victoria had a dinosaur-bearing unit of this thickness been found. Most of the fossils discovered continued to be single bones, but two instances occurred where small groups of bone fragments appeared to be clustered together. This clustering hinted that several bones from a single individual might be together in this thick rock unit. In the Victorian Cretaceous such association of terrestrial vertebrates' bones had only been previously seen at the Slippery Rock site.

The other puzzle about this site was what the increased thickness of the fossiliferous unit meant. Had we been following an ancient stream channel that had passed into a pond or lake? Or could it be that previously we had been excavating the shallow edge of a stream channel and were now approaching the center of that stream? What other reason could there be for the abrupt downward increase in thickness of the fossiliferous unit? If there had been such a depression on the bottom of the ancient body of water, could that depression have been a place where bones frequently came to rest, producing an abundance of fossils never before encountered at a Victorian dinosaur site?

All the crew were thinking about these things as the work continued. Two in particular discussed it endlessly at night—Nina Herrmann, a Danish geology and vertebrate paleontology graduate student, and Nick van Klaveren.

In the middle of the dig, Nick had to return to his professional responsibilities in Western Australia. Nina and three others—Natalie

Schroeder (a member of Dinosaur Cove crews since 1986), David Pickering, and Ivan Kobiolke—formed an enthusiastic band that continued the work.

After several weeks, the high tides began to sweep sand and rocks back into the excavations daily. After much experimentation, a system of sandbag walls held in place with iron rods and steel mesh finally prevented the sand and rock from filling the excavation, but they could do nothing about flooding. Not only did water come over the top of the sand bags but innumerable cracks in the rock of the shore platform allowed the water to flood in. As attempts were made to seal the known cracks, water spurted through new cracks.

Despite the difficulties, the work went on, and when the field season came to an end, there was no sign that the fossiliferous unit was near exhaustion. As it was traced farther west, it continued to become thicker, and fossil bones, if anything, became slightly more common.

Despite the fact that we had had a memorable "Extinction of Dinosaur Cove" party a month before and the National Geographic Society had made it clear that they wished for us to move on to other areas, we could not bring ourselves to walk away from Dinosaur Cove forever with this development unfolding as the field season came to a close. Fortunately, just at this stage, John Lahey of *The Age* newspaper had become vitally interested in the project and had not only written stories about it but arranged for us to write a four-page lift-out for the paper summarizing the entire project. He was able to persuade the management of *The Age* to fund another season of work at Dinosaur Cove.

Clearly the problems of working Dinosaur Cove East would become more difficult if the fossiliferous unit continued in the direction indicated. It was dipping downward at 15° in the direction of the ocean. If it continued in that direction another 10 meters (33 feet), instead of 1 meter (3 feet) of overburden there would be 3½ meters (11½ feet) of overburden. And instead of the ocean merely sweeping around the site frequently at high tide, it would be permanently underwater.

The Twin Excavations of 1994 and Beyond

We wanted to return to Dinosaur Cove East in 1994 because of the unexpected discovery of a tantalizing thickening of the fossiliferous rock there at the end of the 1993 field season. But we had a desire to return to the Flat Rocks site in 1994 as well in order to follow up on the initial work we had done there in 1992 rather than delay work there for another year. Fortunately, it was possible to run the two digs simultaneously because there were sufficient volunteers to do both, and, in particular, there were two people who could lead the respective digs and were really eager to see the work go forward at each of these sites. They were Lesley Kool, who organized and ran the operation at Flat Rocks, and Nina Herrmann, who did the same at Dinosaur Cove East.

This left us free to be able to assist either excavation should unexpected difficulties be encountered. Our faith in these two leaders was fully justified, for both carried out their programs successfully without our help on site.

Nina and her crew of three first cleared out 60 tonnes (66 tons) of boulders and sand that had accumulated in the excavation at Dinosaur Cove East over the winter. They then continued to follow the fossiliferous rock layer downward, ever downward. By the end of the field season, they had reached a depth of 3 meters (10 feet) below the surface of the shore platform—3 meters below sea level. To dig that hole, this small group of people jack-hammered out and shifted over 60 tonnes (66 tons) of brick-hard sandstone. Three pumps were eventually needed to keep the sea out sufficiently for work to go ahead at all. While fossil bones did turn up, the crew did not find the hoped-for concentration in the deepest part of the fossiliferous layer. The number of fossil bones they encountered per square meter was about the same as it had been at shallower depths. The principal difference was that whereas before crews had removed 10–20 cm (4–8 inches) of overburden to recover fossils at Dinosaur Cove East, in the last stages of work there, 3 meters (10 feet) of rock had to be taken off while the sea poured in through the cracks in the walls of the excavation. In addition, what they found was no different from what had been recovered at shallower depths. Because of all these things, but particularly because little new was being found, the end of this field season there really was the end of our work at that locality.

Fortunately, Lesley and her crew found that the fossiliferous rock at the Flat Rocks site continued to consistently produce fossils. This has remained the case in the 1995 to 2000 seasons. While on site, Lesley divided her crew into two groups: the ones who actually excavated the rocks and those who broke them up to look for bones and teeth within each piece. The excavation crew has always been led by Nick van Klaveren, now a very experienced miner with extensive experience in the gold fields of Western Australia.

Lesley and Nick's work is not only producing an older suite of fossils than those at Dinosaur Cove but a different one as well. After ten field seasons at Dinosaur Cove, for example, only two teeth of small carnivorous dinosaurs had ever been recovered. At the Flat Rocks locality, on the other hand, more than 30 such teeth were found in the first two years. Overall, preliminary studies indicate that generally the same types of dinosaurs occur at the two localities, but different genera and species are found in each. This is the sort of change to be expected in a community of animals evolving in an area over a time period of 10 million years when there is only a slight physical change. There is no sign, for example, of a catastrophic extinction event such as the impact of a large extraterrestrial object like the one that may have brought about the extinction of the dinosaurs 65 million years ago. Nor is there any indication yet of a relatively minor change, such as an influx of new groups into the area from elsewhere.

9

Other Eggs, Other Baskets

Because we have only a tiny area, 4 square kilometers (1½ square miles), in which to search for dinosaurs in Victoria, we are well aware that in our own lifetimes it is conceivable that there might be little left to find out about these fascinating animals in that region. Therefore, we have always been alert to the possibility of working in other areas. One of the potential leads we followed up was a find that had apparently been made in Australia a century and a half earlier.

In 1844, four British ships headed by the *H.M.S. Fly* were sent to build a beacon on Raine Island off the eastern side of the Cape York Peninsula. Cape York is the northernmost point on the Australian continent. The beacon was to help ships navigate through Torres Strait as they traveled between the cities of southeastern Australia and those of the southeast Asian region. All four ships remained in the area for four and a half months while the beacon was under construction. During this time the scientists on board carried out a detailed exploration of much of the eastern coastal region of the Cape York Peninsula.

Nearly half a century later, the noted British dinosaur specialist Harry Govier Seeley described a few limb and toe bones of a new dinosaur, to which he gave the name *Agrosaurus macgillivrayi*.[44] In accordance with the rules for naming dinosaurs and other animals, these bones became the holotype of that particular species of dinosaur.

The only locality information Seeley could find was a lone label, "in a small, delicate handwriting, 'Fly, 1844. Jn. Macgillivray, from the N.E. coast of Australia.'" John MacGillivray was a naturalist attached to the Fly Expedition. His job was to collect animals for the 13th Earl of Derby. Unfortunately, his notebooks for the expedition have never been found, and he never mentioned anything about having discovered such a specimen in the book he wrote about an expedition to the same area a few years later, when he sailed aboard the *H.M.S. Rattlesnake*.[45] A geologist, J. Bette Jukes, did publish a book on the results of the 1844 voyage of the *H.M.S. Fly,* but he also never mentioned anything about the discovery of fossil vertebrate bones during the course of that trip.[46] When he described the fossils in 1891, Seeley found they matched

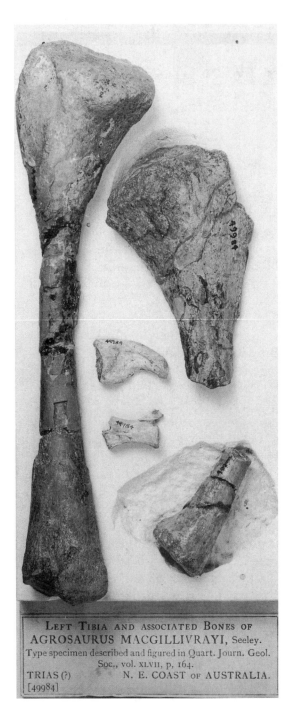

LEFT TIBIA AND ASSOCIATED BONES OF
AGROSAURUS MACGILLIVRAYI, Seeley.
Type specimen described and figured in Quart. Journ. Geol.
Soc., vol. XLVII, p, 164.
TRIAS (?)   N. E. COAST OF AUSTRALIA.
[49984]

Figure 53. The bones of the holotype of *Agrosaurus macgillivrayi.* From left: left tibia, ungual phalanx, distal caudal vertebra, proximal part of right tibia, distal end of right radius. On the facing page is the original handwritten label associated with those fossils. It reads, "By Jn.[Mr?] Macgillivray. From N.E. coast of Australia. 'Fly' 1844." Perhaps the note was originally written for a plant specimen or lizard and got mixed up with the fossil bones later. Whoever caused that to happen certainly led us on a merry chase. And perhaps they also inadvertently set in motion the chain of events that will eventually lead to the discovery of fossil vertebrates on the Cape York Peninsula which have nothing to do with the holotype of *A. macgillivrayi.*

closely with dinosaurs he knew from the Late Triassic and Early Jurassic, about 200 million years ago, early in the age of dinosaurs. Modern opinion concurs. *Agrosaurus macgillivrayi* is classed as a prosauropod, a group of primitive dinosaurs confined to that time.

Fifteen years after Seeley first described *Agrosaurus macgillivrayi*, another renowned student of dinosaurs, the German Frederick von Huene, re-examined and re-described these fossils.[47] In passing, he noted how similar the rock in which the fossil was preserved was to that from an English locality near Bristol called Durdham Down. Durdham Down yielded a dinosaur so similar to *Agrosaurus macgillivrayi* that von Huene decided both belonged in the same genus, *Thecodontosaurus*. Thus, he referred to the holotype bones of *Agrosaurus macgillivrayi* not as such but as *Thecodontosaurus macgillivrayi*.

The classification of organisms has a box-within-a-box structure, the smallest box being the species, the next size up being the genus. Above that are numerous categories that are ever more comprehensive. The most commonly used categories in order of increasing scope are species, genus, family, order, class, phylum, and kingdom. Therefore, von Huene's action suggests that despite the fact that *Agrosaurus* (or *Thecodontosaurus*) *macgillivrayi* is Australian, he regarded it as quite similar to the Durdham Down *Thecodontosaurus antiquus*.

It is intriguing that despite his observations about the similarity in both the fossils and the rock, von Huene never questioned the authenticity of the locality information associated with the holotype of *Agrosaurus macgillivrayi*. But because there was little information available in 1906 about the geology of that remote region of Australia, nothing was known that would have prompted such doubts.

Late Triassic and Early Jurassic continental deposits are not rare in eastern Australia. But for the most part, they are sandstones that have been chemically leached over a lengthy period of weathering. This process usually destroys any fossil bones that once were in them. The one possible exception is the partial skeleton of the primitive sauropod *Rhoetosaurus brownei* from the Early or Middle Jurassic of southeastern Queensland.

Because we knew where the *H.M.S. Fly* was in 1844 and had examined the available geological information about the northeast coast of Australia, we found one rock unit that stood out as being a likely source for the fossils which Seeley described in 1891. These are the Helby Beds that are exposed on the east coast of the Cape York Peninsula, due west of Raine Island. The ships supporting the construction of the beacon on Raine Island repeatedly journeyed to the mainland to obtain wood and fresh water. Both Macgillivray and Jukes took every opportunity to go ashore to collect specimens and make observations. Clearly this was the most likely place for people operating longboats to have collected a block of rock with fossil bones in it.

Not only are the Helby Beds exposed on the coast, but their age, as indicated by current studies on fossil pollen, is just right—Early to Middle Jurassic. The coincidence of all this information enticed us to visit the sites of rock outcrop of the Helby Beds. The expedition would attempt to ascertain whether *Agrosaurus macgillivrayi* might have come from these rocks. We also hoped to find an area which would eventually yield another interesting dinosaur assemblage.

The first step in our search was a reconnaissance flight from Cairns to Cape York along the east coast of the Cape York Peninsula in September 1993. Both of us, accompanied by Lesley Kool and a previous student of Pat's, Greg McNamara, then based at James Cook University in Townsville, hopped aboard a small airplane in Cairns and spent the day flying low from there to Horn Island in Torres Strait and then back, more than six hours of straight, low-level flying. The only rocks of the right age (Mesozoic) that did not appear to be deeply weathered were some of the coastal exposures of the Helby Beds north of Cape Grenville. From the air, some of these coastal exposures had an uncannily familiar appearance to us. Had they been green instead of brown, they would have looked just like the dinosaur-bearing rocks we knew so well along coastal Victoria. The seemingly unweathered outcrops could be seen from the air over a distance of 80 kilometers (50 miles) and had a combined outcrop exposure of about 3–5 kilometers (2–3 miles) along this coast. All the rest of the outcrops we saw were clearly weathered beyond a stage where fossils could have been preserved in them.

Our next move was to set foot on these rocks and see if there really were fossils in them. In July 1995, we hired the *El Torito*, a vessel about 20 meters (65 feet) long, for six days of reconnaissance of all possible outcrops of the Helby Beds between Captain Billy Landing and the Tern Cliffs to the north. *El Torito* was perfect for operating in these waters and easily accommodated our entire field party.

Unfortunately, the holotype of *Agrosaurus macgillivrayi* had been completely removed from the rock in which it was found by the time we first became aware of it. All the associated rock had been immersed in acetic acid, which dissolved the cement binding the individual sand grains together, and only a small bit of acid-resistant residue remained. The only hint about what the enclosing rock looked like were von Huene's comments about it, published at the beginning of the century. This meant we really had no detailed idea of what the rock which produced the holotype looked like. That made it much more difficult to hunt systematically for additional fossils.

Joining us in the search for fossils were eight other seasoned fossil hunters with varied backgrounds, all of whom we had worked with before. Their varied backgrounds were important because their different experiences searching for fossils in many parts of the world made it more likely that one of them might be able to more readily spot a fossil preserved in a particular way than the rest of us. We thus hedged our bets as much as possible when we hand-picked this crew. We were backed up by a ship's crew of four, including the captain. All were interested in what we were doing, and they very capably provided all the logistical support we needed. We could concentrate on hunting fossils! What a luxury!

Our field party departed from Cairns on the *Gulf Express* on July 10, 1995. The *Gulf Express* is an offshore-oil-platform supply ship that is used as a coastal freighter; it makes a weekly return voyage to Thursday and Horn Islands near the tip of the Cape York Peninsula. Late at night on July 11th, the field party transferred to the *El Torito* at sea under fortunately calm conditions near Portland Roads. We moved both ourselves and the boxes of collecting equipment and other baggage from one ship to the other without incident, but it was certainly dramatic against an ink-black sky and the glare of blinding floodlights from the two ships. The time of the transfer was the first of several instances when the weather and sea conditions proved unseasonably favorable. Normally in July, southeasterly winds cause high waves off the eastern shore of the Cape York Peninsula. We hardly had a day with more than a ripple. Luck was on our side. Had the waves been there, our search probably would not have been completed during the short six days we had. That time of year, the Austral winter, was the only time to do this job properly. At any other time it would have been difficult to keep a clear head in the heat and humidity while searching for elusive fossils. During what passes for winter on the far north Queensland coast, temperatures are commonly a pleasant 25°C (77°F), with a cool onshore breeze. During the summer, temperatures can soar to 35°C (95°F) or more, with little breeze and oppressive humidity. On the few days that the sea was running high, the trip to the beach from the *El Torito* in the ship's motorized dingy resulted in thorough drenchings. A real advantage of doing this in the tropics was that even though we were wet, we were seldom uncomfortable.

Once on the beach Gerry Kool (the husband of our chief preparator, Lesley), who had army experience in Vietnam, kept a sharp eye out for saltwater crocodiles. He carried a .350 Magnum carbine at the ready so that the prospectors did not have to watch their backs and could concentrate on fossil hunting. The thought of being attacked by a distant relative of one of the animals we were looking for in the Helby Beds was not a pleasant one. Although suspicious noises were heard around one waterhole and footprints and drag marks were observed on two occasions, no crocodiles were actually sighted. We also had to keep in mind the wild water buffalo in the coastal scrub and the sharks that joined the crocodiles in the beautiful blue tropical waters—those we did see. So, despite the beauty, one never was allowed to have a sense of complete security, except perhaps when on board the *El Torito*.

We came ashore at Captain Billy Landing the morning after our nighttime transfer to the *El Torito*, and prospecting began. The rocks we landed on were reminiscent of the coastal outcrops of the dinosaur-bearing units in Victoria which were so familiar to us. Working northward, we examined all the outcrops of the Helby Beds near Captain Billy Landing by the end of the second day without discovering a single fossil bone, despite the similarity of many of the rocks to those that produced dinosaurs in Victoria. But we did find many scraps of fossilized wood. These rocks were the unweathered remains of sand and pebble deposits formed in river channels nearly 200 million years ago.

On the third day and the first half of the fourth day, we hopped ashore to examine the Helby Beds farther north, which from our aerial search had not seemed remarkably different from those at Captain Billy Landing. On the ground they proved to have been extensively chemically weathered. For the most part, these were laterite boulders and kaolinized rocks. This was not the kind of place where fossil bones were going to be found. Laterite is an iron-rich rock type that is formed when everything else, including any fossil bones that might have once been present, has been chemically leached from the original rock. Kaolinized rocks are those for which most of the major original constituents, such as feldspars and other non-quartzose rock fragments (and also fossil bones), have been chemically converted to clays—mainly white kaolin. We had an immediate bad feeling about these rocks, but we prospected them in detail. Still we found no bones.

The scramble over the boulders for hour after hour, searching in detail for small areas that were not chemically altered, did turn up some places where small patches of rock were similar to those near Captain Billy Landing; but again we found no fossil bones of any kind. By noon

Figure 54. Outcrops at Captain Billy Landing. These were the most promising-looking outcrops we found anywhere on the Cape York Peninsula. In the foreground can be seen some examples of slightly ferruginized rock. Elsewhere, it was much more chemically altered than at this site.

on the fourth day, with no fossil bones found anywhere, we decided to return to Captain Billy Landing for a second search to maximize our chances of finding fossil bones. The *El Torito* spent the afternoon retracing its path and arrived off the southernmost exposures of the Helby Beds less than an hour before sunset. Determined to use every available minute of sunlight, we went ashore even this late in the day. We spread out and in quick succession, Pat found two objects that appeared to be bone at locations nearly 50 meters (165 feet) apart. As we passed them among ourselves in the failing light of day and again on board the *El Torito* that night, we could not totally convince ourselves one way or the other that the objects were bone or not, although with time the doubts became weaker.

Next morning we enthusiastically returned to the beach to begin our search where the two possible fossil bone fragments had been found the evening before. On previous days, we had become spoiled because the seas had been unseasonably calm, making our trips between the *El Torito* and the beach almost a humdrum affair. On this morning, however, the seas were running high, and the field party was quite soggy by the time our feet touched terra firma. It did not matter to us, for we reasoned if the two objects found the previous day in such a short time were really fossil bones, more would be found quickly.

That was not to be. Many of us spent the entire day cracking rock at that site, and we did not find another object like either of the two possible fossil bone fragments found the evening before. How could it be that Pat could find two of these specimens, be they fossils or water-worn rock pebbles, so quickly in the few moments of fading light at the end of one day and then not one of us discover another the entire following day? This puzzled us greatly. Disappointed, some of us re-prospected other outcrops of the Helby Beds near Captain Billy Landing for about half the day, again to no avail.

The final day on the *El Torito* was spent reaching Thursday Island. While under way, we took the opportunity to examine the possible fossil bone fragments under full natural sunlight for the first time. There, with the knowledge that nothing more like them had been found, it was harder to convince ourselves that they were, in fact, small pieces of fossil bones. We carefully re-wrapped them for transport to Monash University, where they could be freed from the surrounding rock, examined under a microscope, and chemically tested for a more definitive analysis. However, before they were carefully wrapped up, Tom nearly threw them overboard in frustration because of their ambiguous nature. Fortunately, he restrained himself. That would have been grounds for divorce!

At Monash University, the two possible fossil fragments found in the Helby Beds were examined at length. We considered a number of possibilities. Were they water-worn rock pebbles? If so, there were two odd things about them. First, on both specimens, one smooth surface was a partial cylinder. This smooth surface was truncated by a smooth flat surface. Such a combination is what might be expected in a fragment of bone but it is unlikely in a water-worn rock pebble. Second, on both specimens there was a very rough surface opposite the smooth cylindrical surface, as if the sponge-like textured inside of a bone had been exposed by breakage. Although pebbles with rough, broken surfaces do exist, they are rare in fine-grained sandstones. In this case, these were the only two objects discovered that had this odd combination of smooth and rough surfaces. This pattern would be more common in a bone fragment than in an ordinary rock pebble, and there were millions of the latter in these sediments.

Visual comparison of a bone of the holotype of *Agrosaurus macgillivrayi* with the two possible fossil fragments found near Captain Billy Landing immediately showed us that our specimen and *Agrosaurus macgillivrayi* were quite different in preservation style. The Captain Billy Landing specimens are solid, whereas the inner part of the *Agrosaurus macgillivrayi* bone, the spongiosa, is open and filled with air. Long before we became interested in this problem, all the rock in which the holotype of *Agrosaurus macgillivrayi* was originally embedded had been disaggregated with acetic acid in an experiment. This suggested that it was originally cemented with a calcium carbonate. The leaching quite evident in all outcrops of the Helby Beds we visited had removed any carbonate that might once have been present in those rocks, and when we dropped acetic acid on our specimens it had no effect.

The difference in preservation between the holotype of *Agrosaurus macgillivrayi* and the two possible bone fragments from Captain Billy Landing, coupled with the absence of carbonates in the Helby Beds, strongly suggested to us that the holotype specimen came from somewhere other than the Helby Beds of the Cape York Peninsula. The lack of other coastal outcrops in "N. E. Australia" of the right geological age that might have yielded a prosauropod to a ship's landing party operating inland on foot in 1844 makes it unlikely that *A. macgillivrayi* came from that area at all.

The rock which once held the holotype of *Agrosaurus macgillivrayi* was described by von Huene as follows: "Das Gestein erinnert in höchstem Grade an die Knochen breccie von Durdham Down in Bristol." [The rock is extremely reminiscent of the bone breccia at Durdham Down near Bristol.] Not only is the rock similar to that at Bristol, but the bones of the holotype are outwardly quite similar to those from Durdham Down, and both are equally and radically different from the two possible bone fragments from Cape York. Analysis of trace elements in all of these specimens shows the same pattern of similarity and difference; chemically the bones of *Agrosaurus macgillivrayi* and those from Durdham Down are similar to each other, and both are quite unlike the Cape York Peninsula finds. This suggests a possible source for the holotype, but it doesn't answer the question of how it came to be labeled as Seeley found it to be when he wrote his paper describing the specimen.

Another similarity is that fossils of the same species of a much smaller animal, *Diphyodontosaurus avonis*, akin to the living tuatara of New Zealand, have been found both in the rock associated with the holotype

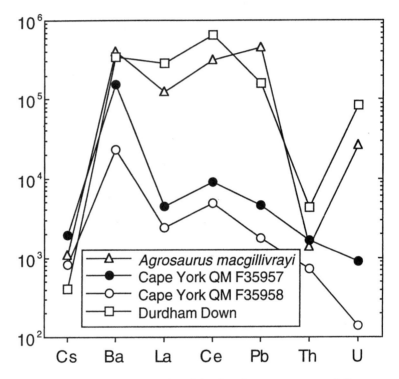

Figure 55. Logarithmic plot of the abundances in parts per billion of seven trace elements in four samples. Note how the abundances in the fossil bone from the holotype of *Agrosaurus macgillivrayi* and the sample from Durdham Down closely track one another while the two objects found on Cape York have a quite different trace element abundance signature.

of *Agrosaurus macgillivrayi* and that from Durdham Down. The only other sites where *D. avonis* is known are other Triassic fissures also in the Bristol area like Durdham Down.

So, in the end, although there had been highly suggestive evidence that the holotype of *Agrosaurus macgillivrayi* came from the Cape York Peninsula, it is almost certain that it did not. Despite the information with the specimen and the coincidence of the Helby Beds being of the right age and in the right place to have been the source, all this supported an interpretation that on further examination was found to have been, once again, "neat, plausible . . . and wrong."

However, the discovery of two possible bones in the Helby Beds may turn out to be an example of what is so often important in the practice of science—namely, doing the right thing for the wrong reason. We will readily seize any future opportunity to visit Captain Billy Landing, even if just for a few hours. After all, the quite similar Flat Rocks site was found after many visits to the area that produced not a single bone scrap because the fossiliferous rock was previously covered with sand. Much of the shore platform at Captain Billy Landing was similarly covered at the time of the visits in 1995. Persistence is a word familiar to most paleontologists. Persistence is still very appropriate with regard to these ancient riverine sediments in far north Queensland. Who knows what the future will reveal in them? Only by going back again and again as opportunities present themselves will a discovery be made there, if it is possible to do so.

10

An Unexpected Surprise

At the end of each dig at Inverloch, Lesley Kool has a party for her crew. On Saturday, March 8, 1997, the party for that season was held. Pat remained home with the kids, as often was the case as school became a consuming issue for Tim. Restful weekends were important and late-night parties just did not fit in. Tom, together with John Herman, "The Mayor of Dinoville," did attend, however. As he walked into the proceedings, Tom asked half facetiously, "Have you got any dinosaur jaws or mammal skulls to show me?" Lesley inwardly gulped in surprise but managed to answer truthfully, "No, but there is one specimen you might want to look at . . . but only if you have time."

After some time, a rock was almost casually produced and Tom glanced at it, seeing only a small smear of brown, a tiny bit of fossil bone. When he put it under a microscope, at first all he could see was a bit of nondescript brown fossil bone. After a first glance at it, he thought to himself, "There must be something about this specimen that made Lesley want to show it to me, but what could be possibly interesting about a formless sliver of bone?" With that thought in mind he looked again and let his eyes wander over all the rock in the field of view and then it hit him almost like a punch in the face. There, right before his eyes, was a bit of enamel lying next to the sliver of bone. The bit of enamel was part of a tooth. So that meant that the sliver of bone was a tiny jaw. What made this tiny jaw different was that the tooth was more than a simple spike, such as are found in many fish and reptiles, and it was quite different from the range of tooth types that occur in dinosaurs. This tooth was clearly that of a mammal, and next to it was another tooth, and another, and another. Four teeth were present, and, while three of them had been broken through when the rock containing them and the jaw was broken open, the last molar was intact. When he looked at the other half of the rock which contained more of the jaw, he saw the rest of the teeth.

All this took less time than it took for you to read the previous paragraph. When Tom looked up again from the microscope to comment about the specimen to Lesley, there were about ten eager faces looking at him and a flashbulb went off almost immediately, the first of many over the next few minutes. The specimen had been found just that morning, and all the crew were eager to see Tom's reaction. They were not disappointed. For several minutes, he sat there, stunned, looking at the specimen from time to time and saying to himself variations of "Oh my Lord," and "Oh my God."

Figure 56. The joy of discovery: Nicola Barton holds the tiny rock fragment containing the first mammalian jaw found at Flat Rocks, the holotype of *Ausktribosphenos nyktos*, moments after it was realized how important the specimen was.

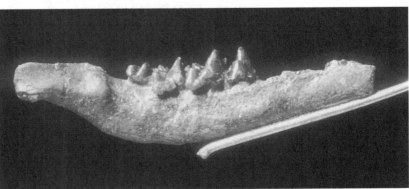

Figure 57. The jaw, a possible placental mammal from the Early Cretaceous Flat Rocks site, Victoria, Australia. (See Plate 15.)

A week later, when the 1997 dig was over, Lesley focused all her attention on preparing the jaw. The specimen was quite clearly much smaller than the only Mesozoic mammal jaws previously found in Australia: those of some monotremes from Lightning Ridge that were about the same age as the fossils from Flat Rocks.[48] However, that was not to say that much smaller monotremes did not exist. Because monotremes were already known to have lived in Australia from rocks of this age, we hoped it would be something else, because it would then represent the first record of another major branch of the mammalian family tree.

Fossil Preparation

Fossils, as they are collected in the field, are seldom in a condition suitable to be studied or displayed. The process of taking a fossil as it is collected and making it available for scientific investigation and exhibition is called preparation.[49] The ways in which fossils are prepared are dictated by the character of the rock in which the specimen is embedded and the nature of the fossil itself. To take an ideal situation, if the individual bones are quite hard and fully intact, having no fractures, and the enclosing matrix is loose sand, the appropriate preparation technique might be as simple as physically removing the bones while recording their positions and orientation. In cases where bones are found in limestone, it is often possible to get exquisite preparation results by dissolving away the rock with acetic acid. This is because the bone is generally still composed of the original calcium phosphate, which is all but completely insoluble in acetic acid. Such preparation can do less damage to specimens than even the finest mechanical preparation of fossils, if the circumstances are right.

Unfortunately, the rock in which the Victorian polar dinosaurs occur is not loose sand, nor can it be dissolved away with any acid that will not also destroy the fossils themselves. Rather, it is brick-hard sandstone, siltstone, and claystone. Furthermore, the bones are softer than the rock around them. To extract the fossils, the rock has to be carefully chipped away. "Carefully chipping away" may conjure up an image of someone with a hammer and chisel patiently tapping away at a fossil. But that happens only rarely. More often, the place of the chisel is taken by a needle held in a handle and the preparator patiently scrapes, grain by grain, with that needle, sharpening it every few minutes. Often, when working with tiny bones, it gets down to teasing away individual grains, first loosening them by carefully prodding with the tip of a needle. A somewhat coarser procedure is to use an engraving tool. In either case, the work is exacting, demanding patience, knowledge, and a steady hand. Fortunately, Lesley Kool combines those attributes with an undaunted enthusiasm for the project.

After being home for only three days, Lesley called late at night and

Tom immediately drove to her house to have a first look at the jaw. It was prepared sufficiently that the unbroken molar was fully visible, and much of the mandible was exposed as well. Unlike some of our colleagues, Tom is not a walking library of detailed information about vertebrate paleontology. As a matter of course, he double-checks every generalization he makes in a scientific paper. So what was so startling to him when he first looked at the fully prepared jaw was that there were three characteristics that pointed to its being a placental mammal that were so basic that he did not have to confirm any of them by looking them up in a book. Those features were sufficient to show that the jaw was neither a monotreme nor a marsupial, the two mammalian groups it seemed most reasonable at the time to expect in the Early Cretaceous Australian fauna. The form of the mandible and the number and shape of the teeth suggested that this specimen might belong to another group of mammals, the placentals. The identification of it as a possible placental was completely unexpected.[50] There were no flashbulbs popping this time but there should have been: thousands of them would have been appropriate for the way Tom felt in response to this totally surprising conclusion! Further preparation of the specimen revealed nothing to counter this initial identification.

The Thrill of Scientific Discovery

There are two different types of scientific discovery that can send a shiver down a paleontologist's spine. The first is finding a rare, beautiful fossil. There is an undeniable satisfaction in finally finding a fossil that you have long sought after. It is the feeling that comes with accomplishing at long last something for which you have put in a considerable amount of effort. The second kind of scientific discovery is more sublime. Perhaps because of that, the satisfaction is longer lasting. This is the moment of insight when you realize the unique significance of a particular fossil and know that you are the first human to appreciate it. Insights of this kind do not happen often. An individual scientist is lucky if it happens once in a lifetime; if it does, it is mind-blowing.

A telling measure of the novelty of an insight is whether it has ever been previously hypothesized. A truly novel insight is one that comes to its creator that no one in their wildest imagination has ever tentatively previously thought of before even as a "What if?" The realization that this fossil, now the holotype of *Ausktribosphenos nyktos*, might be a placental mammal was just such a case. Certainly it had never occurred to us prior to the discovery of *A. nyktos* to consider the possibility that terrestrial placentals might have occurred in Australia during the Early Cretaceous. Before this fossil was found, such an idea was so outlandish that it simply never would have crossed our minds to ask such a question. And we have never heard of anyone else having previously suggested this idea as even a remote possibility. This is a case where nature was stranger than what anyone could imagine. The moment of insight into such an aspect of nature is truly unforgettable.

Once it was fully prepared, Lesley turned the jaw over to us. We then started to describe and analyze this tantalizing specimen. Even before Lesley had completely finished preparing the jaw, Peter Trusler had agreed to prepare scientific illustrations of it. He was actually quite flattered by the request, remarking in the course of the conversation that "I would be honored to do so."

The reason that high-quality photographs of the jaw would not be sufficient on their own to adequately illustrate the fossil was that Lesley could not repair all of the damage done when the specimen was found. It was desirable to present a detailed reconstruction of how this specimen had appeared before the damage was done as best as could be determined, and the only way to do so was to have Peter create a representation. We found it very helpful to have Peter's illustrations before us when looking at the fossil through a microscope. While we could mentally correct for damage to a particular area of one tooth, it was difficult to visualize the whole specimen as it might have looked prior to being damaged. Peter's illustrations made such visualization straightforward. As with the *Diprotodon* skulls he illustrated earlier, when we compared details of Peter's drawings with the fossil, we found complete concordance.

The occurrence of a possible placental in Australia far earlier than any previously known there—and more than twice the age of any marsupial yet discovered on the continent—was so unexpected that we were extremely skeptical of our hypothesis. We kept searching for a flaw in the reasoning which pointed to the jaw's placental affinities. Besides trying to find weaknesses ourselves, we distributed copies of a preliminary description and of Peter's partial illustrations for critical comment to a number of colleagues in South and North America and Europe. Of those, fifteen responded with detailed criticisms and suggestions. Between them, they pointed out a number of weaknesses in our analysis of the fossil. That aided us immeasurably in focusing our attention on particular aspects of the argument. We continued to revise and refine the analysis until, eleven weeks after the specimen was discovered, we sent a manuscript about the jaw to the journal *Science*.

Other Outlooks

All scientists do not analyze similar phenomena in the same way. Just how much this is so is one of the greatest surprises we have had in our professional careers. Such differences are not simply due to the fact that no two people have had the same experiences in life. Nor is it just simply a matter of different data sets. There clearly is not a single scientific method that all scientists follow, although there are many attitudes of mind and approach to problems that most scientists share.

Scientists seek out an explanation for the phenomena they observe. Many have the idea that because there must have been a particular cause, their job is to find the unique, correct explanation. Their procedure is to propose an explanation, and if it fails, to put up another one. All the time, they are seeking the final answer. In their minds, the latest hypothesis for a given phenomenon is the final answer until another one comes along to replace it. If the latest hypothesis is correct, that will never happen. For this reason, such a scientist may express shock when a colleague simultaneously advocates two different explanations for the same phenomena.

Scientists who do propose more than one hypothesis as a possible explanation for a phenomenon are following the tenets of the geologist Thomas Chamberlin, who advocated what he called the Method of Multiple Working Hypotheses.[51] "With this method," he wrote, "the dangers of potential affection for a favorite theory can be circumvented." Certainly if one thinks up as many different explanations as possible for a phenomenon, one avoids the risk of being mesmerized by one's own answer.

Another difference in scientific outlook that was particularly startling to us was the realization that attitudes toward verification of a hypothesis can be quite different. A colleague who had produced an estimate of one of the environmental factors was quite taken aback when we wished to point out in a paper that independent data supported his original interpretation. He quite earnestly felt that it was a slap in the face. In his view, his estimate had been arrived at by beginning with first principles and, therefore, could not be wrong if the work was done correctly. Hence, to verify the work independently was to imply that he might not have been careful. To us, on the other hand, such independent confirmation of a hypothesis is the hoped-for outcome of any idea that is put forward, for no scientific concept has ever been so exhaustively tested that it cannot, in principle, be found to have its limitations, even if it is not downright wrong.

A classic case is Newtonian physics. For more than two centuries it was thought to be complete, fully describing all physical phenomena. Newtonian physics was so successful and widely accepted, in fact, that physics was thought to be a dead field for researchers. But in the twentieth century, it was subsumed by quantum mechanics and the theory of relativity. The development of both quantum mechanics and relativity theory was triggered by observations that could not be explained by the application of Newtonian mechanics. In those cases, Newton's theory was found to be restricted to masses larger than a molecule and velocities much less than that of light, domains in which the theory is quite successfully used to this day. It was not wrong, but it was limited.

Any scientific idea that stands up that well three centuries after it was proposed is a quite fruitful one, even though it is not complete.

Another factor that influences the outlook of scientists is the nature of the particular science they do. The experimental physicist who can keep all conditions constant while changing a single variable has a quite different view of how science is done than a paleontologist who deals with a multitude of factors, many of which are unknown and in principle unknowable. The reality of this is reflected in the fact that the personalities of scientists in different fields are noticeably different. If one were to walk into a gathering of either physicists or paleontologists and did not know which group was present, it would not take long for the differences in their collective personalities to enable the perceptive observer to figure out what fields they studied, even if they were not discussing their work.

But whether a scientist seeks final answers or multiple working hypotheses, one thing is true about them all. If a phenomenon is so well understood that there can be no doubt about what is the best scientific explanation for it, then scientists regard it as no longer interesting except as a stepping-stone for understanding some other phenomenon. The scientific enterprise thrives on uncertainty. What makes it interesting to its practitioners is not the well-established central body of knowledge but the ragged frontier of ignorance.

Our analysis of the jaw could not have been done in such a short time without the aid of modern communication, for we were repeatedly in contact with colleagues in Europe and North and South America. In a number of instances we exchanged e-mail and FAX messages a dozen times with a single individual during this period of intense activity. Had this discovery been made as recently as 1980, this level of interaction with overseas colleagues would not have been possible due to the physical isolation of Australia from so much of the world's scientific community. At most, we might have had one, or with a bit of luck, two exchanges of airmail letters in that time. Certainly, the tempo would have been much different, and the final product would have been much poorer because of the lack of outside input.

Because boat mail would have required at least four months for a reply, if we have been writing in 1935, there would have been no outside input at all in the time between the discovery of the jaw and the time when the article was submitted eleven weeks later. That rapid interaction with other scientists that was possible in 1997 trimmed away defects and oversights that would probably have been otherwise published. Our good fortune to be wrestling with this problem in 1997 constantly came to mind as the number of messages that passed back and forth so readily via these inexpensive, fast, and reliable means of communication neared and then exceeded the one hundred mark.[52]

The writing of this manuscript thus served to sharpen up our ideas about the jaw. It is one thing to have an insight into some phenomenon. It is another to carefully think the whole matter through and weed out misconceptions that might significantly distort its proper interpretation, if not result in a totally erroneous analysis. A true help to that weeding-out process is having to set out your ideas in a logical manner in a manuscript. As an aid to clarifying thinking, this process alone would justify writing of scientific papers, even if the information and conclusions were ultimately communicated in an entirely different manner.

Once the manuscript was received by the editor at *Science,* it was sent out to be read by two anonymous reviewers. This peer review is standard practice at most scientific journals. Both such reviewers disagreed most vehemently with the conclusion that the animal might be a placental mammal. The paper was rejected. Upon receiving the letter of rejection, Tom read the comments superficially and set it aside. After a month, when he had cooled down a bit, he read it again and decided that the objections raised could be addressed. He wrote to the editor at *Science* and received permission from him to answer the points raised by the reviewers. In reply, the editor noted that this was a privilege rarely granted and made it clear there would be no third chance. Tom spent another month writing and polishing an eight-page reply, about the same length and the result of the same amount of writing effort as the original manuscript. He sent it to the editor at *Science* along with a slightly altered version of the original manuscript. In short order, the editor, having checked the eight-page reply with one of the anonymous reviewers, accepted the paper, and it was published 25 weeks after it was first submitted. That was a little more than twice as long as from the moment of discovery to the submission of the initial manuscript.

Publications of papers in scientific journals often takes what can seem to the author to be an inordinate amount of time and effort. With journals such as *Science* and *Nature,* the competition for space is fierce. The majority of manuscripts submitted to both are rejected simply because there is not room to publish so many. There is a strong pressure on editors to weed out weak papers and shorten those that are accepted. Anonymous reviews play a vital role in this process. Reviewers judge whether the paper is worthy of publication in the journal. They may spot significant errors of fact, logic, or presentation and assist the author by bringing such problems to his or her attention. Also, they may point out relevant information unknown to the author that may or may not be already published. When it is implemented fairly, this process of peer review serves the readership of a journal well. This is because the readers know that what they are putting their effort into reading and understanding has been gone over by other experts than the author to at least attempt to weed out errors and confusing writing. So although the experience of publishing this paper in *Science* definitely had its frustrating moments, the procedures encountered served the purposes for which they were put into place reasonably well.

Clearly the jaw represented a previously unknown mammal and hence needed a name. Twice during the process of writing the manuscript for *Science* we altered the name as our thoughts about this specimen changed. It was finally christened *Ausktribosphenos nyktos,* as mentioned above. The generic name *Ausktribosphenos* sacrificed euphony for a highly compressed but conservative interpretation of what the specimen was. *Aus* is the Latin word for "southern" and thus refers to Australia, *K* is the standard geological abbreviation for Cretaceous, the age of the fossil, and *tribosphenos* is a technical way of describing the teeth of this animal. Individual tribosphenic teeth are specialized for both cutting and crushing food in a manner found in primitive marsupials and placentals, but not in the monotremes.

Based on what we now think we know, the jaw could just as well have been named *Auskplacentalia,* "the Australian Cretaceous Placental mammal," one of the names considered earlier but discarded. However, it may well be that, although at the present time all the evidence, as we interpret it, points to *Ausktribosphenos* being a placental mammal, further discoveries could force a re-evaluation of that opinion. That such caution is warranted is borne out by the opinions of the fifteen quite capable colleagues who commented in detail about the specimen. Thirteen would not accept it as a placental, one was uncertain, and one was convinced that it was. In any case, the jaw is surely a new discovery and is tribosphenic, so the name *Ausktribosphenos nyktos* is appropriate.

The specific name *nyktos* means "night" in Greek and was chosen to refer to the prolonged nights this species would have experienced during winter because southeastern Australia was within the Antarctic Circle during the Early Cretaceous. And it is not incidental that the word pays homage to three of the people associated with the discovery of this specimen: Nicola, Nicholas, and Nicole. It was Nicola Barton, in fact, who had spotted the jaw while cracking rock on the morning of the party. Thinking it might be particularly interesting, she showed it to Lesley, who first set it aside while she finished processing other specimens. Only when Lesley turned her full attention to it did she realize that here was something really different. Because it was quite unlike any dinosaur or fish jaw she had ever seen, she thought it must be a mammal. She was not sure, though, because her familiarity with Mesozoic mammals was from pictures rather than from firsthand experience. But by the time she had passed it around among the crew, she and they were fully convinced that it was a mammal. And so they laid the ambush for Tom to walk into that evening, camera and all at the ready.

A year after the first specimen of *Ausktribosphenos nyktos* was found,

Lesley had to move some of the many boxes filled with rocks from the Flat Rocks site that had one or more fragments of bone showing in each of them. This was to protect the fossils from bushfires. Simply because she had all this material in front of her, she decided to re-examine some of it under a microscope she had at her home. She had last seen these specimens while peering at them through a hand lens in the field. As she worked through this material, she noticed a number of interesting fossils. She continued to search through this material from time to time as opportunities presented themselves between other jobs. She persisted until she was suddenly taken aback: there was a second mammal jaw that also had four teeth preserved.

When Tom had a chance to look at it, he confirmed it was either *Ausktribosphenos nyktos* or another closely related but as-yet-unnamed species in the same family. Most important of all, it had the lower part of the back of the jaw preserved. There a feature, the mandibular angle, was preserved. On the first specimen found of *A. nyktos*, the holotype, this area had been so heavily, but smoothly, abraded that Tom did not recognize the small part of the mandibular angle that was in fact preserved on that jaw. The apparent absence of that structure was one of the most serious objections raised against the theory that *A. nyktos* was a placental.[53] With the discovery of the second specimen with an angle, it did not occur to Tom to go back and have a second, closer look at the holotype to see if in fact there was the tiniest remnant of it preserved. He just assumed that the difference between the two specimens was because of individual variation in this feature. The suggestion to do so came in an e-mail from renowned Polish vertebrate paleontologist and explorer Zofia Kielan-Jaworowska. Only then did Tom see it. That it should have been she who suggested this was quite ironic because she, together with Richard Cifelli and Luo Zhexi, had made the case that *A. nyktos* was not a placental based in part on the reported absence of the angle in the holotype.[54]

With two such jaws now known from the site and about 100 boxes of rocks containing bone scraps stored away to be re-examined, Lesley forged ahead. In the next two months, she found two more small mammal jaws. One was just a toothless scrap, complete enough only to be identified as a probable mammal. The other, however, contained one worn tooth, and was initially interpreted by us as belonging to an entirely different group of mammals, the eupantotheres.

Eupantotheres are thought to be the group out of which both the marsupials and placentals, and quite possibly the monotremes, arose. They had evolved by the Jurassic and were previously known to have lived on all continents except Australia and Antarctica. Unlike the discovery of *Ausktribosphenos nyktos* in the Australian Early Cretaceous, the discovery of a eupantothere in the same age rocks at the same site,

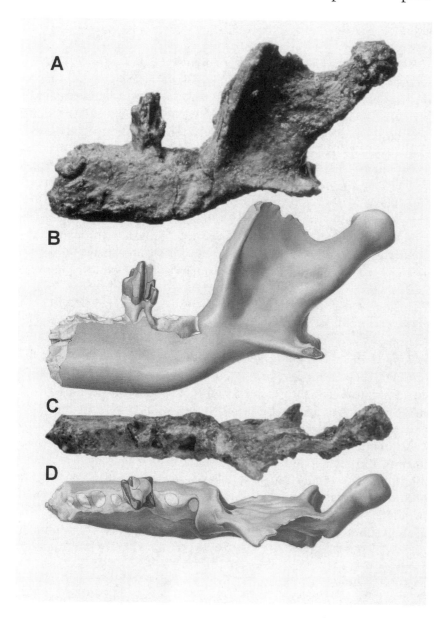

Figure 58. Comparison of photographs and paintings of the same two views of the holotype of *Teinolophos trusleri* provides a graphic example of why an artist of Peter Trusler's skill is vital for the scientific study of many of the fossils we encounter. Even after hours of meticulous preparation by Lesley Kool, the specimen remained heavily encrusted with rock that she could not remove without risking the destruction of the fossil in the process. Therefore, the only way to reconstruct its appearance was to have Peter remove the encrusting rock in his mind and paint the image. A. Photograph of the outside of the jaw. B. Painting of the same view as A. C. Photograph looking down on the jaw. D. Painting of the same view as C. *Photographer: S. Morton.*

although none had been found on the continent before, was not so surprising.

The new mammal was given the name *Teinolophos trusleri*. *Teinolophos* means "the extended loph" in Greek. It refers to the presence at the back of the one known molar of a crest, the likes of which had never been seen in any eupantothere. *Trusleri* was in honor of the artist Peter Trusler. It was fitting that Peter painted a technical illustration of "his" fossil just as he had of the first specimen of *Ausktribosphenos nyktos*.

Because of the extremely controversial nature of *Ausktribosphenos nyktos*, it was highly desirable to enable those with specialized knowledge about Mesozoic mammals to have casts of the jaw. Besides being an outstanding vertebrate paleontologist, Don Russell is one of the most well-qualified people to make a mold of a delicate fossil.[55] We found this out by chance in 1983. At that time, Tom had just published a paper on a tiny mammal jaw from China that was about the same size as *A. nyktos* with Prof. Minchen Chow, then director of the Institute of Vertebrate Paleontology and Paleoanthropology in Beijing.[56] After the paper was published, Minchen asked Tom to have technicians at the National Museum of Victoria mold the specimen. Having spent hours looking at the specimen under a microscope, Tom was aghast at the prospect. The technicians who were being asked to do the job were equally aghast. They were convinced that the delicate teeth were so shot through with micro-fractures that they would break up if molding compound were placed on them and removed after setting.

His request denied, Minchen said nothing. But within a year he flew to Paris, and there Don Russell took on the task of molding that incredibly delicate specimen. Later, Minchen returned to Australia, bringing both the original specimen and casts of it with him. When we compared the two under a microscope and with the illustrations in the publication about it, it was clear that Don Russell had done the seemingly impossible; he had molded the specimen with no damage to it and obtained a perfect cast.

Tom was flying to Europe eighteen months after the holotype of *Ausktribosphenos nyktos* was found and could have easily stopped by in Paris much as Minchen had done fifteen years earlier and there Don Russell could have molded the specimen. However, after more than a quarter of a century of looking for such a fossil and because at the time it was the only one known from the Victorian Cretaceous, we would not risk it. Rather, we asked Don to fly from Paris to Australia to mold this one specimen.

About the time Don was boarding the plane in Paris, the second mammal jaw was found. When he arrived in Australia, Don set to work molding the first jaw while the second one was being prepared. Because he had to return to Paris after a set time, he decided to mold what was exposed of the second jaw even though it was not yet completely out of the rock. It was well that he did. Shortly after the mould was removed, the rear half of the most posterior molar broke up as it was being prepared. Lesley was almost in tears. However, because Don had fortuitously molded the specimen just hours before, a record of the tooth existed. Although it was still heart-wrenching for Lesley to have the fossil break up on her, knowing that a cast could probably be made of it consoled her somewhat. Don took the mold with him to Paris, where an experienced technician at the Muséum National d'Histoire Naturelle, Philippe Richir, poured the first cast in the delicate mold. Needless to say, we were relieved when that cast arrived in Australia without having been lost in transit.

And what of the birds that, along with the mammals, had been our primary objective for two decades as we searched the Mesozoic of Australia? They, too, have put in an appearance at the Flat Rocks locality. Unlike the first mammal from there, the first bird specimen was not regarded as a bird when it was first seen in the rock. But when it was prepared out of the rock, it had a shape that was unmistakably that of a wishbone of a bird, which has not yet been identified. The wishbone was found just three days before the first mammal was recovered from Flat Rocks.

An Hypothesis Tested

When she has reached the stage where her experience tells her that to remove any more matrix from around a fossil would endanger it, Lesley has the good sense to stop. One of the mammals found during the 1999 field season required just such treatment. It had two teeth, and for all we could tell, it could either have been one of the two species already known from Flat Rocks or something completely new. But how to get it out?

Thirty years before, we had spent two summers in southeastern Wyoming collecting Late Jurassic mammals from Como Bluff, the first site in North America to produce both a large quantity of dinosaur bones and Mesozoic mammals.[57] One of the members of the crew was Charles Schaff, then of Yale University and now based at Harvard. Chuck is a highly skilled preparator, and we asked him to come to Australia for a week to help Lesley out. He came and taught Lesley much about preparation techniques and also made many suggestions to improve the setup in her preparation facility. He found the preparation of the specimen which was the reason for his trip to Australia in the first place so difficult that at times he needed a break from working on it. Chuck's idea of a "break" took the form of removing the matrix from the lone tooth on the holotype of *Teinolophos trusleri*. When he finished the

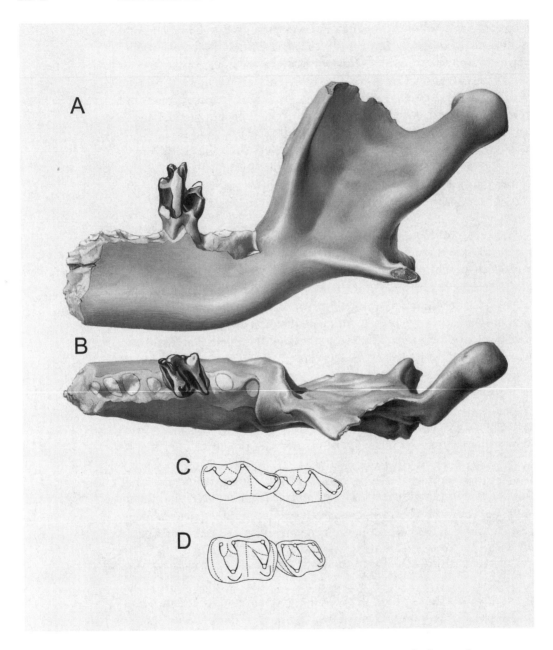

Figure 59. The revised illustrations of the holotype of *Teinolophos trusleri.* Compare with Fig. 58. A. Painting of the outside of the jaw. B. Painting looking down on the jaw. C. Occlusal view of lower molars of the eupantothere *Peramus tenuirostris,*° which is what we expected *T. trusleri* to look like. D. Occlusal view of lower molars of the monotreme *Steropodon galmani,*° which have a closer resemblance to the lone tooth of *T. trusleri.*

°After Kielan-Jaworowska, Crompton, and Jenkins (1987).

job, we were in for a real surprise. We were astonished because instead of having the kind of tooth pattern we expected for a eupantothere, it looked remarkably similar to that of the undoubted monotreme *Steropodon galmani* from Lightning Ridge, except that it is only about one-fifth as big.

Every identification of a fossil is a scientific hypothesis open to challenge. Our revision of what we think *Teinolophos trusleri* is, is a perfect example of that. The original identification of it as a eupantothere was well founded on the basis of the evidence available to us at the time. When we identified *T. trusleri* as a eupantothere, we had a general expectation of what was under the matrix, which Lesley dared not remove. When Chuck's preparation of that fossil revealed something quite different, we had to think again. No described monotreme has the rear part of its jaw similar to that of *T. trusleri* and all eupantotheres. In particular, monotremes all lack a highly prominent mandibular angle, the same structure which played a key role in the early debate concerning whether or not *Ausktribosphenos nyktos* was a placental. However, there is a catch to that statement about monotreme jaws, because the only ones for which that posterior area is known are Late Cenozoic and modern species. It is quite conceivable that when the back of the jaw of *S. galmani* is discovered, it will resemble *T. trusleri*. But, in the meantime, what to make of *T. trusleri?* Is it a monotreme with a jaw like a eupantothere or a eupantothere with a tooth like a monotreme? Because monotremes are thought to have arisen from eupantotheres, that such an intermediate form might be found is, in retrospect, no surprise. The important difference between the features allying *T. trusleri* with the eupantotheres on the one hand and the monotremes on the other, is that the tooth structure which resembles *S. galmani* is an advanced feature, whereas the resemblance of the jaw to those of eupantotheres is a primitive feature. Therefore, *T. trusleri* can be regarded as a monotreme for much the same reason that *Archaeopteryx lithographica* is regarded as a bird. In both cases, the two species can be seen as intermediates between two major groups, a primitive group that gave rise to a more advanced group. Because both species share at least one advanced feature with the derivative group, in each case they are assigned to the advanced group rather than the more primitive one. This is because they are hypothesized to share an advanced feature or features with the oldest species belonging to the derivative group because they are descendants of that species which is the group's common ancestor. The common ancestor of the derivative groups could either be those species themselves (*Archaeopteryx lithographica* and *Teinolophos trusleri*) or, far more likely, a species that is as yet undiscovered (and it probably never will be).

If *Teinolophos trusleri* is correctly allocated to the monotremes, its unusually small size for a monotreme is intriguing. Here may be yet another instance where Darlington's hypothesis that there was a radiation of monotremes of which we know only a few disparate lineages is corroborated. In any case, with *T. trusleri* evidently somehow allied to the monotremes, the "extended loph" on the molar which had been the basis for the generic name *Teinolophos* does not now appear unusual at all. But once a scientific name is given for a species and genus it is fixed even if the original rationale for the name evaporates, as this one apparently has.

11

Getting through the Winter

Anusuya Chinsamy studies the microstructure of fossil bones.[58] She and Pat met in 1993 at the Society of Vertebrate Paleontology annual meeting in Albuquerque, New Mexico. As a result of that meeting, Anusuya began an investigation into the histology of the polar dinosaur bones from southeastern Australia.

In order for her to do her work, Anusuya must grind a thin section of a fossil bone to be studied. Because there is so little dinosaur bone from southeastern Australia, we were reluctant to sacrifice any of the specimens that could be identified to the generic or specific level of refinement. After much soul-searching, we sent her a fragment of *Timimus hermani*, the ornithomimosaur, and a hypsilophodontid unidentified as to genus but definitely not *Leaellynasaura*, although both specimens came from Dinosaur Cove.

Anusuya completed her study and sent the results to us. After a quick glance at them, we set them aside because of other pressing matters. Later, after carefully re-reading her manuscript, we immediately made a connection with an interpretation that we had published almost a decade earlier concerning the enlarged optic lobes of the hypsilophodontid *Leaellynasaura*. Anusuya had noted that whereas the ornithomimosaur *Timimus* showed lines of arrested growth (LAGs), the hypsilophodontid from Dinosaur Cove did not.

LAGs are dark bands in bone formed when the rate of deposition of the bone crystals is reduced. This happens when a vertebrate significantly lowers its metabolic rate for some reason. Such a lowering of the metabolic rate can happen because of lack of food or water, other environmental stress, or a period of hibernation. The absence of LAGs in the hypsilophodontid implied that it never experienced any of these, including, most intriguingly, hibernation. That a hypsilophodontid from Dinosaur Cove did not hibernate added powerful support to the idea that these animals were active during the polar winter. The previous evidence was that *Leaellynasaura* had enlarged optic lobes, which implied to us that it would have had the ability to see under the extremely low light conditions that prevailed during the depths of the polar winter.

That the hypsilophodontid from Dinosaur Cove did not have LAGs turned out to be part of a more general pattern seen in these small ornithischians. Anusuya had found a lack of LAGs in three different dinosaurs in this family from lower paleolatitudes, whereas most other dinosaurs consistently had them, as did *Timimus*. Although this suggested that *Leaellynasaura*, too, would not have LAGs, the sample on which this pattern had been demonstrated was too small for us to be confident without checking a specimen of *Leaellynasaura*. We carefully repeated the agonizing process of sacrificing part of one of the few *Leaellynasaura* femora we had. Before doing so, the specimen was molded and cast. Then a few cubic millimeters were cut out and the gap filled with plastic. In the same package also went a sample from another as-yet-unnamed hypsilophodontid from the Flat Rocks locality.

Frozen Ground

Andrew Constantine is an imaginative and meticulous scientist. Over the years he has built up, in painstaking detail, a picture of the various ways in which the sands and muds were deposited that were to become the rocks where our dinosaur fossils occur. Bones rarely were buried and subsequently fossilized on the floodplains far from the main channels of the streams that cut across the broad rift valley that was formed as Australia and Antarctica began to move apart. Instead, most of the dinosaur bones accumulated in newly formed channels of either braided or meandering streams. As the name implies, a braided stream has several channels that flow for a distance and then meet only to split up again farther downstream. They form a pattern of lines that intertwine like a schoolgirl's braided hair. In contrast, meandering streams have a single main channel that typically crosses the floodplain in a snake-like or sinuous pattern. Over time the meandering channel shifts back and forth, cutting new channels and abandoning old ones. The beginning of a new channel typically happens during a flood, when a stream bursts its banks and water gushes across the countryside, cutting into the existing flood plain. Sediments laid down during this initial phase have a typical appearance that is labeled by geologists who study them "crevasse splay deposits." Most of the dinosaur bones found in Victoria were found in these sorts of rocks.

In working this out, Andrew has carried out many related studies. For example, he has reviewed in detail the fossil pollen evidence for the age of all the coastal outcrops of the Strzelecki Group, so it is now possible for us to know immediately the age of any fossil we find there. This followed on the fossil pollen work done by Barbara Wagstaff and Jennifer McEwen-Mason. One thing that Andrew could not explain for a long time was a peculiar structure in some of these rocks. In the sandstone below a thick claystone unit near our Flat Rocks locale, he found instances where there were lumps of the overlying claystone. How could the overlying claystone come to have been incorporated into the rock unit below it? One of the guiding principles of sedimentary geology is known as Steno's Law of Superposition. It states that unless a sedimentary rock sequence has been overturned by subsequent events, the younger rocks overlie those which are older. Yet here the younger rocks were somehow inserted into the older ones. What had happened to explain this seeming violation of Steno's Law of Superposition?

Figure 60. The peculiar outcrop which puzzled Andrew Constantine for a long time. The shading outlines the area in which the clay has evidently sunk into the sand. For an explanation of what circumstances may have brought this about, see Fig. 61.

In search of a solution to this problem, Andrew spent long hours reading geological literature in areas far removed from those customarily consulted by persons with an interest in Mesozoic sedimentology. Eventually he came across books on structures formed in modern times in glacial regions that closely resembled what he had found in the Strzelecki Group.

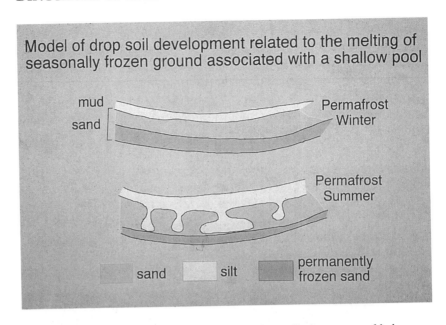

Figure 61. Formation of a drop soil structure in seasonally frozen ground below a shallow pool. The top diagram shows a cross-section of a frozen shallow pool. The top layer is a compact clay laid down at the bottom of the pool. Beneath it are two layers of frozen sand, perhaps as much as 50% water by volume. In the lower diagram, the upper layer of frozen sand has melted to become a slurry in the summertime. The clay above, having less water in it than the slurry, is denser and, therefore, tends to sink into the sand-water slurry until it reaches the still frozen, deeper layer of water-saturated sand. When the clay reaches the still frozen sand, it can no longer sink any deeper and tends to flatten out against its upper surface. From Vandenberghe (1988) (used with permission).

In the modern examples, the counterpart of the underlying sandstone is a frozen mix of sand and water that is often more than 50 percent water by weight. It might be best described as a frozen slurry. It forms in shallow depressions. On top of that a soil develops which corresponds to the overlying claystone. In regions today with a mean annual temperature between –2°C (28°F) and +3°C (37°F), the underlying slurries typically melt during the summer to a depth of 10–20 cm (4–8 inches). Because the compact clay has much less water in it, the overlying soil is denser than the slurry of sand and liquid water beneath it. Therefore, there is a tendency for the soil to sink into the slurry until it reaches the still frozen slurry. At that point, the bottom of the tear-shaped chunks of soft soil tend to flatten out against the harder, permanently frozen slurry below—that is, a permafrost horizon. Imagine removing the water from such a sequence of slurry and clay and turning it all to rock and you have just what Andrew found.

The temperature range of −2°C to +3°C is close to the mean annual temperatures of Fairbanks, Alaska (−3°C or 27°F) and Qiqihar, northeastern China (+3°C) (see Fig. 30). This mean annual temperature estimate is similar to what Bob Gregory had deduced a decade before when analyzing carbonate concretions. Because the most prominent of these sequences was just 3 meters (10 feet) below the Flat Rocks site, we became very interested in learning what Anusuya would discover about the hypsilophodontids found there.

An Unexpected Synthesis

Anusuya's results for *Leaellynasaura* were just the same as those of the other, unnamed Dinosaur Cove hypsilophodontid that she had initially examined: no lines of arrested growth (LAGs). The hypsilophodontid bone from Flat Rocks looked quite different. Yet here too there were no LAGs. The difference was not a reflection of something that had happened in life but of what had happened to the bone long after it was buried. A few hundred meters away from the Flat Rocks locality there is a body of igneous rock that had intruded into the sediments about 90 million years ago. Such a rock body is called a dike. The heat from that then molten rock raised the temperature of the surrounding sedimentary rock so much that the microscopic structure of the bone was altered. This baking caused the difference that Anusuya saw.

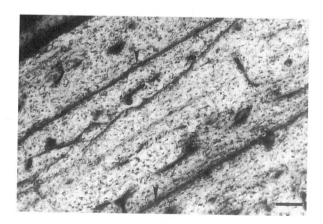

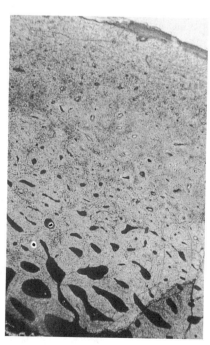

Figure 62A. *Above.* Cross-section of a femur of *Timimus hermani* clearly showing the lines of arrested growth, or LAGs. *Photographer: A. Chinsamy.*

Figure 62B. *Right.* Cross-section of a femur of a hypsilophodontid with not the slightest indication of the presence of LAGs. *Photographer: A. Chinsamy.*

Andrei Sher is a Russian geologist from St. Petersburg who is interested in understanding the climatic evolution of the last million years, when the most recent episode of Ice Ages occurred. He is also very interested in the fossils of mammals that lived during those times, especially the miniature mammoths that survived until 3,700 years ago on Wrangel Island in the Arctic Ocean just north of eastern Siberia. Because of these combined interests, he has an extensive firsthand knowledge of Arctic geology. These interests led him to attend some vertebrate paleontological meetings in Perth, Western Australia. Because he was visiting Australia for the first time, he took the opportunity to travel to the eastern Australian states. He stopped at Monash University for a visit and to present a seminar.

When Andrei learned of the presence of sediments nearby that possibly had been frozen ground during the Cretaceous, he was anxious to see them. We arranged for him and Andrew Constantine to go to the sites together. As soon as he saw the evidence, he was persuaded that their origin was likely due to freezing and thawing, just as Andrew had thought. The one thing that he found difficult to get accustomed to was that these 115-million-year-old rocks were dry and brick-hard, quite unlike the frozen ground he knew in present-day Siberia.

When he returned to Russia, Andrei showed photographs of the sites to a colleague, Nikolai Romanovskii, who is based in Moscow and has extensive experience analyzing permafrost deposits. Andrew had been conservative in interpreting the structures as evidence for seasonally frozen ground. He hesitated to infer that beneath the structures that indicated the former presence of ground that froze and thawed each year there had been true permafrost present, that is, ground which never melted throughout the year. He hesitated because there are many places today where seasonally frozen ground does occur with no permafrost beneath it. Andrew had noted that the tear-shaped soil blobs that descended into the underlying slurry had flattened bottoms, as if they had settled on a frozen surface. That was enough to convince Nikolai Romanovskii, who could only look at photographs, that yes, indeed, there had been permafrost beneath the seasonally frozen ground.

Because these unique insights into the paleoclimate of Early Cretaceous southeastern Australia and the behavior of the dinosaurs from there all came together at about the same time, they were combined into a single paper which was rejected by both *Science* and *Nature* as "not being of general interest." So we turned to the southern hemisphere's answer to those two journals, *The South African Journal of Science,* which published the paper as a cover story. We felt some empathy with Wegener.[59]

Subsequently, the National Geographic Society flew Nikolai Romanovskii to Australia specifically to examine at first hand the evidence for the former presence of permafrost. He not only confirmed that indeed permafrost conditions had prevailed, but he also recognized evidence for another phenomenon indicative of cold climates: patterned ground. Patterned ground is formed when moist clays freeze and contract, forming cracks. If the cracks become filled with sand, then when the clay thaws and expands again, the sand acts as a wedge, buckling the clay. These buckled areas typically form irregular polygons when seen from above, hence the name "patterned ground."

Just as Andrew had previously been the first to recognize the signs of seasonally frozen ground anywhere on Earth during the Mesozoic Era, these observations of evidence that permafrost and patterned ground had been present in Victoria were likewise the first for that era anywhere on the planet. The picture of dinosaurs that were active in a country with permafrost, and that consequently experienced winter temperatures which only a few mammals tolerate today, is a very different one than the popular perception of all dinosaurs as denizens of steaming, tropical swamps.

What we are presently trying to establish is whether the dinosaurs, the other vertebrates, and the flora we have found were living at the same time the permafrost existed. That they were close in time, from the perspective of today, is undoubted. But if the three meters of rock that separates the evidence for permafrost from the dinosaurs at the Flat Rocks site took 20,000 years to be laid down, conditions could possibly have changed significantly in that time. Twenty thousand years ago much of northern Europe and North America was under a thick sheet of ice that is no longer there. There is evidence that major climatic swings did occur in the Mesozoic elsewhere in that amount of time. One of the problems with the permafrost evidence is that the conditions have to be just right for it to be preserved. It may, therefore, have been much more widespread than the available evidence suggests. The concretions necessary to measure oxygen isotope ratios, on the other hand, are much more common. The oxygen isotope measurements do not show signs of major, quick changes in the climate. This suggests that the dinosaurs probably did experience the harsh climatic conditions necessary to have produced permafrost and patterned ground. But the question is far from settled.

So Many Hypsilophodontids

While these studies of the physiology and pathology of the hypsilophodontids were under way, what was constantly at the back of our minds was another question that we were less successful at addressing. This was the at first sight seemingly more straightforward question of just how many different kinds of hypsilophodontids there were in polar southeastern Australia during the late Early Cretaceous. We knew from having written our first paper about them, which was published in 1989, that answering this query by analyzing isolated bones and teeth of these animals was fraught with difficulty.

However, as the collections grew in size, it became ever more apparent that there were many differences among the femora of these dinosaurs. Because of their apparent adaptability to high latitude environments, the idea that the Victorian hypsilophodontids might be more diverse than their lower latitude cousins (where they are usually only a minor component of the dinosaur assemblage) was something almost to be expected. But expecting is one thing; demonstrating with fossil evidence is quite another. So we really wanted to answer this fundamental question about their diversity in our high paleolatitude localities.

We had been fortunate over the years to study a number of collections of femora of hypsilophodontids and other small bipedal ornithischian dinosaurs in museums around the world, which helped us gauge the significance of the variation in Victorian hypsilophodontid femora. What disappointed us about most of the collections we studied was that while such femora did exist in these collections, the sample sizes were surprisingly small if we restricted our studies to femora that were more than 90 percent complete.

One collection which we knew about that was long inaccessible to us was at the Humboldt Museum in Berlin. This was a sample of femora of *Dysalotosaurus lettow-vorbecki* [*Dryosaurus lettow-vorbecki*]. When Wolf-Dieter Heinrich showed the collection of those fossils to Tom in 1998, Tom was almost stunned. There before him were fifteen exquisitely preserved femora of this single species of small bipedal ornithischian from Quarry IG at the famous Tanzanian Late Jurassic dinosaur locality of Tendaguru. These were far more such femora from one place, all belonging to a single species, than we had seen anywhere else. All the fossil bones, not just the femora, found in Quarry IG belong to just the one species, *D. lettow-vorbecki*. These little dinosaurs were thought to have died during widespread drought conditions in the Late Jurassic.[60] The femora varied in size by a factor of three. Thus, it was an ideal sample from which to assess the degree of variation typical of the femora of a single species of small bipedal ornithischian dinosaur. The most vivid impression that Tom came away with was how little difference there was in proportions between the largest and smallest of these femora.

When he returned to Australia and re-examined the Victorian hypsilophodontid femora in light of what he had seen in Berlin, Tom could only conclude that there were probably at least half a dozen hypsilophodontids represented in Victoria. But when the time came to assign individual femora to a particular group, the path ahead suddenly seemed less certain. Unlike the Tendaguru femora, there were usually only two to four femora in any single group. Furthermore, crushing as a consequence of deep burial (up to 3 kilometers [2 miles] at one stage in their history) made it often difficult to determine whether the differences we see now between two Victorian hypsilophodontid femora had existed when the dinosaurs were alive. There was certainly the possibility that something happened after their deaths that might have altered the shape and proportions of the bones, making these differences of no consequence as far as the identification of the femora was concerned.

The end result of this re-study of the diversity of the Victorian hypsilophodontids in 1998 and 1999 was that the answer to our original question of diversity based on femora seemed no closer to being answered than it did a decade before. But what did come out of our re-study was the recognition that the hypsilophodontid dentitions, although paradoxically much more alike, did show evidence for a genus that was completely unrepresented in the collections a decade earlier. Three jaws had been found in a single season at the Flat Rocks locality, jaws that were shorter and deeper than in any other hypsilophodontid known. The best of these became the holotype or basis for the name of

A

Figure 63. Left jaw of *Qantassaurus intrepidus.* This is the holotype of *Q. intrepidus,* that is, it is the specimen on which that name is based. A. Outside view. B. Inside view. Length of the jaw is 56 mm (2⅕ inches). *Photographer: S. Morton.*

B

a new hypsilophodontid, *Qantassaurus intrepidus* (see Fig. 63). The generic name was in honor of QANTAS, the Australian airline which had long assisted us with dinosaur research and exhibitions.

The holotype was found by Mrs. Nicole Evered. Nicole is a longtime resident of Inverloch who was delighted when our operations shifted to her neighborhood. She has helped out with the digging every year since the beginning of work near Inverloch, both on the shore platform and in arranging for needed local support.

Where Did They Come From and Where Did They Go?

One of the questions that naturally comes to mind about the dinosaurs of southeastern Australia is "Where did they come from?" Questions of this nature belong to the field of paleobiogeography. This question is put this way partly because everything does have an origin somewhere. But in the way this question is framed there is an unstated assumption that dinosaurs came from elsewhere rather than assuming that they were the descendants of a lineage with a long history in Australia. This assumption arises because dinosaur fossils are rare in Australia and so many are found elsewhere, particularly in east Asia, western North America, and southern South America. People get the idea that fossils are found where important events in the history of the groups represented by those fossils must have happened in the past. Undoubtedly, the living animals did exist in those places where many fossils can be found today. However, those localities now rich in fossils may have had little relation to where the animals originated or how many were alive in various places at various times. If no rocks were laid down in an area when the dinosaurs were alive, a record of dinosaurs will never be found there no matter how hard anyone searches for them; but this does not mean that they never lived there.

A big stumbling block to pinning down the place of origin and particularly the time of entry of ancestral stocks of Australian dinosaurs is that there is precious little evidence of the group on the continent prior to the late Early Cretaceous (115 million years ago). That is when the specimens we have of the oldest Victorian dinosaurs were part of the skeletons of living animals. Their ancestors could have lived for a long time in Australia, but their prior presence is only documented, if at all, by footprints. Again, it is not that dinosaurs did not exist then in Australia, it is just that their bones either were not preserved or that no one has found them yet. Fossilization, particularly of land animals, is a highly unlikely event. If the geological circumstances prevailing where animals lived were not favorable for eventual burial and subsequent fossilization, they could have thrived without leaving a single trace. Even if their bones were preserved in rock, unless the subsequent geological processes were also favorable (such as uplift and recent exposure), traces of the fossils might not have become visible and thus "show themselves" to the paleontologist searching for such clues about the past during the few centuries that the science of paleontology has been pursued.

From the dinosaurs that we do know, it appears that Australia was not as isolated in the Early Cretaceous as it is now and has been during the Cenozoic. As we explained in Chapter 1, with the possible exception of a single placental from the Eocene of Queensland, during most of the Cenozoic (the past 65 million years), all the known terrestrial mammals in Australia were monotremes and marsupials. Then about 5 million years ago, rodents, which are placental mammals, reached the continent from the north by island-hopping across the Malay Archipelago. There is no known dinosaur as uniquely Australian as the marsupial koala is a uniquely Australian mammal. While the Australian dinosaurs for the most part have been assigned to their own genera, they can be allocated to sub-orders, if not families, known elsewhere. When dinosaurs arose in the Late Triassic, about 210 million years ago, the continents were still combined or had recently been joined in a single major landmass, Pangea. As the Jurassic passed, fragmentation of the continents began. In the initial stage of separation, the continents divided into the southern set plus India (a mega-continent called Gondwana or Gondwanaland), and the northern set, Laurasia. As this process continued, the pathways of potential interchange between continents presumably became more difficult to cross. However, only in the Late Cretaceous did strong regional differences start to appear among the terrestrial vertebrate faunas of the various continents.

With the current state of knowledge, Victorian dinosaurs can be divided into two groups in terms of their paleobiogeographic history. The first are those that were established elsewhere well before the beginning of the Cretaceous. In this first group are the hypsilophodontids, ankylosaurs, and large theropods (including the allosaurids). The small theropods possibly belong here as well. During the Jurassic, Australia was in essence a large peninsula of eastern Gondwana, and faunal movement could have happened in either direction. Presumably the immediate link with the rest of Gondwana would have been through India, which was attached to what is now the western coast of Australia. Alternatively, the immediate link for Australia may have been through East Antarctica, which was joined to it along the southern Australian coast.

The Ankylosauria, or armored dinosaurs, appeared in the Middle Jurassic and are known to have been on all the continents, including Antarctica. Their Antarctic and South American records are in the latest Cretaceous, whereas their time of appearance on other continents is much earlier. The ankylosaurs are one of the best known Australian

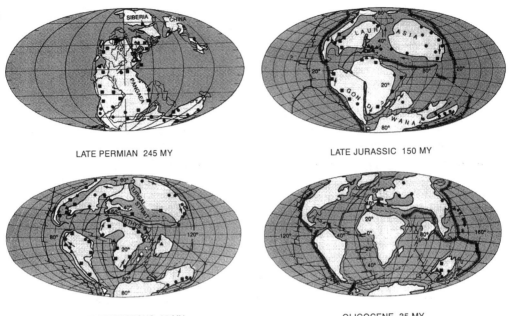

LATE PERMIAN 245 MY

LATE JURASSIC 150 MY

LATE CRETACEOUS 65 MY

OLIGOCENE 35 MY

Figure 64. Paleogeographic reconstructions of the Earth from the late Paleozoic to the mid-Cenozoic (A. Late Permian. B. Late Jurassic. C. Late Cretaceous. D. Mid-Cenozoic). Major divisions of the Earth's crust are shown: hatched lines represent trenches; heavy black lines, ridges; and light black lines, transform faults. Indicators of past climates are plotted on these maps: black squares are salt deposits indicating desert conditions; black triangles are reefs, today restricted to within 30° either side of the Equator, that is, the tropics; and black circles are coals, which form under humid conditions, usually in cool temperate to temperate latitudes. (Permission of L. Frakes and P. Vickers-Rich.)

dinosaurs, for, unlike many other groups which have a record on the continent based on the most meager of evidence, the genus *Minmi* (Fig. 50) is known from three partial skeletons as well as several less complete specimens found in Queensland. As with several other groups, there is just enough fossil material known from Victoria to indicate that the ankylosaurs were present: a few isolated teeth, rib fragments, and wedge-shaped pieces of bone that were buried in the skin.

The second group of Victorian dinosaurs are those whose appearance in Australia is either the earliest record for the group, or at least among the earliest occurrences. That there should be such groups could mean that they did originate in Australia. Or it may be no more than a reflection of the fact that the Early Cretaceous record of dinosaurs worldwide is much more meager than the Late Cretaceous one and that Australia has its best record of dinosaurs during this time of generally poor representation. In this second group are the protoceratopsians, ornithomimosaurs, oviraptorosaurs, and possibly some of the other small theropods.

Oviraptorosaurs are small dinosaurs which, although toothless, are members of the Theropoda. They are best known from the Late Cretaceous of eastern Asia, although they are rare even there. Two jaws of that age are also known from North America. There is also a possible occurrence of a North American oviraptorosaur from the late Early Cretaceous, a record approximately contemporaneous with the tentative one from Australia. Archosaurs such as crocodiles and all dinosaurs have a hole in the jaw, the mandibular fenestra. This hole in oviraptorosaurs is an unusually large one. Because of the large size of the mandibular fenestra, the shapes of the bones surrounding it are highly characteristic, and it was this feature which made it possible for Phil Currie[61] to recognize a fragmentary jaw from Australia as likely belonging to this group.[62] The skull of these animals is even more unusual than the jaw; it is a single rigid unit filled with air spaces that has a pronounced downward-projecting ridge of bone on the roof of the mouth. Some species had a casque similar to that of a cassowary. It is uncertain what the function of this feature was.

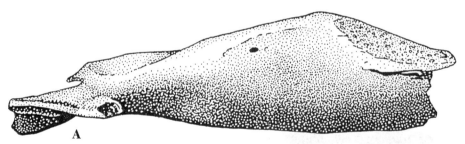

A

B

Figure 65. This jaw fragment and vertebra (see also Plate 16) are the scanty evidence on which the presence in Australia of the oviraptorosaurs is based. The jaw fragment (A) is the primary evidence. The vertebra (B) could possibly belong to some other theropod, but there is no evidence for any other theropods of this size from the same locality. *Photographer: S. Morton.*

While it may well be the case that these groups have their earliest records in Australia merely because of the generally poor Early Cretaceous record of dinosaurs elsewhere, there is another intriguing possibility. China and southeast Asia were formed by the fusion of a series of micro-continental plates, or terranes, which had their origin in the part of eastern Gondwana to the north and northwest of Australia.[63] During much of the Phanerozoic, successive terranes broke away from eastern Gondwana and moved northward to eventually collide with and accrete to southeastern Asia. Figure 66 shows the Late Jurassic–Late Cretaceous phase of this process.

Plate tectonic reconstructions of the entire Earth for any stage during the Early Cretaceous, such as that in Figure 64, commonly show nothing but open ocean between Australia and southeastern Asia, a gap half again as great as the shortest distance between modern-day South America and Africa. By contrast, the reconstruction in Figure 66 that focuses on this area shows several probable islands in that late Mesozoic ocean that now form part of the southeast Asian landmass. While these islands may have existed, it is unlikely that island-hopping during the Early Cretaceous would have been sufficient by itself to enable interchange of terrestrial mammals and dinosaurs between Australia and Asia via that route. If Cretaceous island-hopping were feasible, it is difficult to understand why the reasonably well-known middle Cenozoic terrestrial mammalian faunas of Asia and Australia are so extremely different—they do not share a single family in common. Presumably such an island-hopping route would have become progressively easier for terrestrial mammals to traverse as Australia approached Asia during the Cenozoic.

However, there is still another possibility—a combination of hopping from island to island and being carried northward on one or more of the micro-continents. Figure 66 shows that individual landmasses required about 80 million years to make the transit from eastern Gondwana to the Eurasian landmass. For example, in the Late Jurassic, the West Burma micro-plate was reconstructed as part of the eastern Gondwana landmass. At the end of the Cretaceous, it was part of the Eurasian landmass. Hence, the West Burma micro-plate may have carried some of the Australian dinosaurs and mammals to Asia on its back.

If 80 million years was required for such a northward transit from Australia to Asia, it would appear to have been a rather slow boat. However, as a micro-plate first broke away from Australia and again as it approached Asia, the opportunity for dispersal first onto and then off such a micro-plate would have been increasingly likely. But in this context, it is important to realize that no dinosaurs have been recovered in Australia from Cretaceous rocks older than those we have been explor-

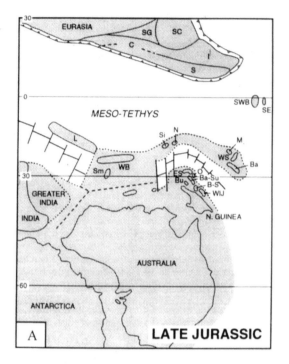

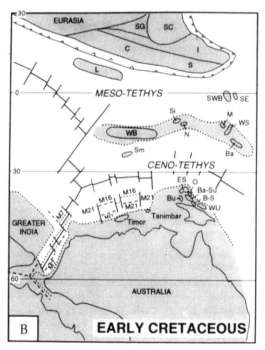

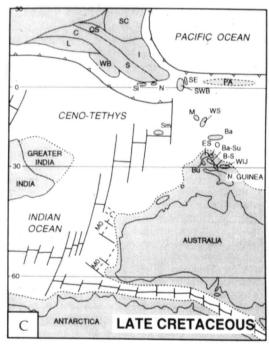

Figure 66. Paleogeographic reconstructions for southeastern Asia and northern Australia, (a) Late Jurassic, (b) Early Cretaceous, (c) Late Cretaceous. A possible mechanism to transport terrestrial vertebrates from Australia to southeast Asia may have been a natural Noah's Ark. Specifically, it may have taken the form of a small landmass or micro-plate that became detached from some area of eastern Gondwana which lay to the north or northwest of what is today Australia and drifted northward to become part of Asia 80 million years later. For example, the western part of modern Burma was one of these micro-plates. It is identified as WB on the three successive maps. In the Late Jurassic, it was part of Gondwana. In the Early Cretaceous, it was an island between Australia and southeast Asia. By the Late Cretaceous, it had become part of southeast Asia. Modified from Fig. 16 in Metcalfe (1996), used with permission of Blackwell Scientific Publications, Melbourne, Australia.

ing in the southeast. There is every possibility that an Australian dinosaur stock that is similar to what we know but tens of millions of years older may have been on a micro-plate that subsequently docked in southeast Asia in the Late Cretaceous.

Such a mechanism would not only explain the apparently early appearance of dinosaurs in Australia but may solve a problem with the mammal *Ausktribosphenos nyktos* as well. There are features about *A. nyktos* which suggest it is not only a placental but, more particularly, is a primitive relative of the living hedgehogs, the Erinaceidae. In that case, the timing of the first appearance of the Erinaceidae in the northern hemisphere 50 million years after *A. nyktos* lived in Australia makes sense, using McKenna's "Noah's Ark" model.[64]

Such a route might also explain other distributional anomalies. South America does not have records of any of the following dinosaurs prior to the last stage of the Cretaceous, if at all: ornithomimosaurs, oviraptorosaurs, protoceratopsians, and ankylosaurs. This is at least 30 million years after they are first known in Australia. All of these four groups are well known in Asia by the early Late Cretaceous. So, a direct connection between Asia and Australia would explain why these dinosaur groups are present on both those continents and did not appear in South America until much later, if ever. Because the micro-continents only moved north, if such a connection between Australia and Asia did take place, it was a one-way trip.

Some Outrageous Speculation

Where were the ancestors of humans when dinosaurs were alive? The only sure thing to say is "somewhere on the Earth's dry surface." But there is a provocative, if rather fragile, combination of observations which suggest that it may be possible to be more specific. Namely, eastern Gondwana might have been their home at that time.

There are four parts to this hypothesis, two of which were previously discussed. First, assume for the moment that *Ausktribosphenos nyktos* does have something to do with the hedgehogs. Likewise, assume that the dispersal of Early Cretaceous hedgehogs from Australia and elsewhere in eastern Gondwana by the mechanism of terranes that rifted away from the northern edge of the Australia–New Guinea plate did take place. The third part of the argument is that the oldest known fossils attributed to the Primates, the mammalian order to which humans belong, are in the northern hemisphere and are Paleocene in age, as is the next oldest erinaceid after *Ausktribosphenos nyktos*.[65] The fourth and final part of the argument is based on comparison between species of the structure of their molecules of DNA and RNA by molecular biologists. Such evidence weakly suggests that there may be a particularly close relationship within the placentals between hedgehogs and primates.

A number of conflicting hypotheses regarding the higher relationships among placental mammals have recently been put forward by molecular biologists. These dissimilarities are based in part on differences in the particular species examined and the disparate parts of the DNA and RNA analyzed in various studies. Thus, as might be anticipated, the suggestion of a close relationship between hedgehogs and primates based on this type of data is far from a universal view. In fact, the molecular evidence for a special relationship between primates and hedgehogs is certainly the weakest link in the argument that eastern Gondwana *just might* have been the home of our ancestors when the dinosaurs lived. The following recent statement is a typical overview of the molecular results: "The hedgehog was in all cases on its own, ambiguously placed in various positions in the [family] tree, depending on method of analysis."[66] But some workers have noted a strong clustering together of the hedgehogs, primates, and rodents.[67] What does seem to be consistent is that neither hedgehogs nor primates consistently fall into one of the major super-ordinal groupings that are commonly recognized on the basis of molecular evidence. One of these, for example, is the Afrotheria, a cluster of six orders that today are all found in Africa and seem to have originated there.[68]

An Early Cretaceous primate or proto-primate found alongside *Ausktribosphenos nyktos* at the Flat Rocks site would be the strongest support imaginable for the idea that our ancestors did dwell in eastern Gondwana when the dinosaurs ruled the Earth. While such a discovery would certainly be highly persuasive, it cannot be realistically demanded as a minimum condition for acceptance of this theory. Even if the theory is true, the likelihood of such a discovery is low. If this idea has any merit, it is probably more reasonable to expect a strengthening of the first and fourth parts of the argument that initially suggested this possibility. Likewise, if either of those lines of evidence is ever shown to be incorrect, the support for this idea would vanish.

With regard to the first part of the argument, the affinities of *Ausktribosphenos nyktos* with the hedgehog, the resolution of that question is probably to come with the discovery of additional fossils. While subtle features of the lower dentition of hedgehogs distinguish them from other mammals, it is the upper dentition, particularly the most anterior molar, that is most distinctive. Unfortunately, thus far, only lower dentitions of *A. nyktos* have been found. That this is the case is the same reason that Samson slew the thousand Philistines with the jawbone rather than the skull of an ass. Lower jaws are much more durable elements than skulls. Despite this, with enough further work at the Flat Rocks locality, an upper dentition of *A. nyktos* will almost certainly be found.

Establishing the degree of affinity between the hedgehogs and primates on the family tree of placental mammals is something that will probably be resolved by the ongoing work of molecular biologists now active in the field.

And the Work Continues

By the start of the 1998 field season at the Inverloch site, what had been a minor problem when work there began had slowly grown into a major headache. As the incoming waters swept over the site with each tide, the excavation was filled with sand. As the hole that was dug to obtain the fossils became steadily larger and deeper, the quantity of wet sand to be dug out at the beginning of each day grew steadily as well. By the end of the 1997 dig, two hours were needed to complete this tiresome task of shifting about 5 tonnes of wet sand before the work of excavating the fossiliferous rock could begin. Something had to be done.

On his own initiative, Nick van Klaveren began to search around for a solution. The most obvious way to proceed was to construct a wall around the site. This was rejected because it would have been an eyesore in such a highly public area. Also, it would have needed to be jackhammered apart at the end of each field season. Even then there would have been traces of its base that would have been an unsightly scar on the shore platform. Eventually, working together with Pat O'Neill, Nick came up with a plan that seemed extremely simple and elegant: fill the $2\frac{1}{2}$ cubic meter ($3\frac{1}{4}$ cubic yard) hole with something that could readily be removed the following day to displace the sand.

Putting a simple idea like that into practice, in fact, took much time and effort both before and during the 1998 dig. As Nick and Pat O'Neill thought it through in detail, a hundred and one practical problems arose and were solved. And that was before the dig even started. The objects chosen to displace the sand were plastic containers. They are hollow and light and are easy to remove, weighing practically nothing compared to the wet sand. But they also float. A method had to be found to keep them underwater during high tide. Vertical steel rods were anchored to the rock. After the plastic containers were in place, a network of I-beams and steel mesh was attached to the vertical rods to hold the plastic containers in place. When this was first tried, it was found that sand that was carried in in a high-velocity stream of water would become wedged into the spaces between the plastic containers and the surrounding rock, lifting the containers upward and filling the hole.

Wrapping the whole mass of plastic containers in tarps eventually solved that problem but significantly prolonged the process of filling the hole at the end of each working day. Even with this added feature, the time required daily to remove the I-beams, steel mesh, plastic containers, and tarps from the hole and later to replace them at the end of the day was much less than two hours by the end of the 1998 season.

Nick had learned much and was able to refine the successful procedures substantially by the time the 1999 dig began. Equally important, when the work was finished at the end of the 1998 and 1999 field seasons, there was not a trace of the equipment left or any visible marks on the rocks of the shore platform itself to show that anything had been done at the site. The tide, which caused the problem in the first place, had obliterated all traces that any work had been done at the site a few hours after the last crew member left the beach at the end of those digs, as it always will.

Figure 67. The excavation site at Inverloch. Because we found the specimens in the open and did not have to tunnel, the work could go ahead much faster than at Dinosaur Cove. However, the problem of keeping sand out of the site continually taxed the ingenuity and patience of the field crew working there. Comparing the quite different problems of excavating the two sites, we, at least, were quite happy with the trade-off in working at Flat Rocks. There we know that except for the admittedly ever-present danger of incoming meteors, none of the crew is likely to be fatally struck by a falling rock while digging!

12

Where Are We Now; Where Are We Going?

The picture that has emerged over the past two decades of the polar dinosaurs of southeastern Australia, the flora and fauna associated with them, and their physical environment during the Early Cretaceous between 100 million and 120 million years ago has come together in an erratic fashion. As you followed the evolution of our ideas, you were presented with a series of partial images, much as the way they became known to us at the time. But the overall picture may well have been missed in the twistings and turnings as our ideas developed together with all the minutiae presented. Therefore, in this chapter we would like to present you with the overall picture of what we have learned as it appears to us at present.

Between 100 and 120 million years ago, Australia was far to the south of its present location, joined to Antarctica, which was then, as now, close to the South Pole. Southeastern Australia lay inside the Antarctic Circle and then experienced between one-and-a-half and four-and-a-half months of continuous darkness and continuous daylight each year. The exact figure depends on precisely where within the Antarctic Circle Australia was located.

The rift valley between Australia and Antarctica originated during the Jurassic Period as those two continents began to separate from one another. This was the start of the last major division of the Gondwana super-continent, which had begun to break up 50 million years earlier. Prior to that, Pangea, the even larger super-continent incorporating all the major landmasses on Earth, had separated into Laurasia in the north and Gondwana in the south. During the Early Cretaceous the movement of Australia away from Antarctica was slow, only a few millimeters per year, much less than the approximately 100 millimeters per year that it is at present and has been for the last 45 million years.

As the two continents pulled apart, the floor of the rift valley sank, but at the same time there was a vast outpouring of volcanic rock from volcanoes probably located to the east near the Lord Howe Rise. The ash from these volcanoes filled this sinking valley. After the ash fell to the ground, it was repeatedly reworked by the water coursing down major rivers. Those westward flowing rivers emptied into the eastwardly encroaching sea. That sea came from the west because the separation between Australia and Antarctica was a scissors-like motion, with the pivot point located near the southeastern corner of Australia. The sea was to arrive in the area where the dinosaurs have been found about 10 million years after they had lived.

There are no deposits formed of ash particles that fell from the sky and were never transported again. After reaching the ground, all the ash was moved about by rivers but that did not go on long enough for those particles to weather much. As a consequence, feldspar minerals that are unstable on the Earth's surface are a major component of these rocks. This reflects the fact that the outpouring of the ash was rapid; burial was quick.

The deposition of sediment in the rift valley lasted for about 25 million years. As a result, the fossil vertebrate sites preserved in that ancient valley span a significant period of geological time, but not all 25 million years. Sites in the Otway Ranges to the west of Melbourne date to about 105 million years, while those in the Strzelecki Ranges are about 115 million years old.

Numerous tributaries flowed into the major rivers. As these tributaries meandered back and forth across the floor of the rift valley, creating a floodplain, they constantly reworked the sediments. It is at the bottom of these tributaries where sands and clay gall clasts accumulated that the occasional fossil bones are preserved. On the floor of the floodplain, away from the tributaries, water flowed more slowly. For the most part, only clay was deposited there, for the current was too weak to bring sands or clay gall clasts and fossil bones into that environment. The extremely rare fossil bones found in such regions belonged to animals that died close to where their remains have been discovered. During floods, tributaries would sometimes burst their banks and cut new channels. In doing so, they would pick up the debris in their path, which included the rare bone as well as dried-up pieces of clay and plant matter. That which could not float tended to settle in the bottom of the newly formed channels and accumulated there in areas where the flow of water was slowest, such as the downstream side of a sandbar. It is that process of sweeping the countryside of debris which accounts for the fossils found in the Early Cretaceous rocks of southeastern Australia.

Most of the fossil sites were located well away from the edge of the rift valley. An area that may have been significantly closer to the edge was one that favored the preservation of labyrinthodonts. These amphibians are found in the coarsest sandstones and conglomerates that occur in the Strzelecki Ranges. The coarseness is a reflection of deposition from fast-running streams, which occurred on the steeper slopes near the shoulders of the rift valley.

When the fossil accumulations in the Strzelecki Ranges were forming, the area was at times bitterly cold. These rocks bear definite signs of permafrost, the seasonally frozen ground and patterned ground that are features characteristic of high latitudes today. Evidence for seasonally frozen ground has also been recognized in the Otway Ranges. But we do not find evidence of permafrost and patterned ground in the Otways, perhaps because that region was somewhat warmer than the temperatures that prevailed when rocks were formed in the Strzelecki Ranges 10 million years before.

It is uncertain whether dinosaurs, other vertebrates, and plants lived in southeastern Australia when the conditions were cold enough during the Early Cretaceous to form permafrost, patterned ground, and seasonally frozen ground. It is quite possible that when the harshest conditions prevailed, the immediate ancestors of the plants and animals we know as fossils were living in warmer parts of the continent and only re-entered southeastern Australia when the climate became somewhat milder.

The flora that lived alongside the dinosaurs during the Early Cretaceous in southeastern Australia was dominated by green, for flowering plants were either extremely rare or non-existent. While fungi, mosses, liverworts, ferns, and cycads were present, the prominent elements were the extinct seed ferns (pteridosperms), the nearly extinct scouring rushes (sphenopsids), the evergreen conifers, and the deciduous ginkgoes. What sets this flora apart from modern polar floras is the greater size of the largest trees.

Any meaningful analysis of the terrestrial vertebrate fossils from the Early Cretaceous of southeastern Australia must always take into account that there is a pronounced size bias in the sample in favor of individuals that ranged in size from shrews to people. This does not mean that larger animals were not present, only that their fossil remains were seldom preserved. Occasionally, the smallest pieces that can possibly be identified of large dinosaurs, such as ankylosaurs and theropods, have been recovered. All we can learn from them is that such larger forms were there, but at least we know that. This size bias exists because the small streams in which the vast majority of our fossils were found were simply incapable of moving large bones to where they could be accumulated, concentrated, and buried.

The latest Cretaceous terrestrial vertebrate assemblage from the

North Slope of Alaska is quite similar to contemporaneous ones much farther south in Montana and Wyoming. Then, as now, those more southern areas were much warmer. While the fauna is similar, there is one conspicuous difference. Alaska lacks groups such as lizards and snakes, forms that today are known to be cold-blooded animals. From this, Clemens and Nelms concluded that all the Alaskan terrestrial vertebrates, including the dinosaurs, were likely to have been warm blooded and thus to have adapted to a polar environment.[69] The same generalization applies to the Early Cretaceous vertebrate assemblages of Victoria, for lizards and snakes are conspicuous by their absence, although animals as small or smaller do occur there at that time.

Although most hypsilophodontid species from southeastern Australia are known from only isolated femora or teeth, their great diversity and number suggest that polar conditions particularly suited them. Furthermore, the hypsilophodontids were apparently well adapted for the polar environment. Their big eyes and large optic lobes made it possible for them to see under conditions of low light and be active even in the depths of winter. Their ever-growing bones suggest they never shut down, hinting at warm-bloodedness. Other dinosaurs, such as the ornithomimosaur *Timimus hermani*, probably coped with the winter conditions by hibernating. Their bones show lines of arrested growth (LAGs), indicating that they did shut down in times of stress each year.

Compared with other vertebrate faunas of the same age elsewhere, the Early Cretaceous polar fauna of southeastern Australia is an odd mix of relics from the past and advanced forms for their time. Holdovers from the Jurassic include an allosauroid and a labyrinthodont amphibian as well as a fish more typical of the Triassic, the palaeonisciform *Psilicthys*. Advanced forms include early, if not the earliest, records of ornithomimosaur, protoceratopsian, and possibly oviraptorosaur dinosaurs and a placental mammal that may be allied with the hedgehogs. Other dinosaurs in southeastern Australia, such as the diverse hypsilophodontids, are typical of the Early Cretaceous while still others, such as the small theropods, are so poorly known that there is not enough evidence to say whether they are at all unusual in this regard.

Surprisingly, in light of the apparent proximity of South America via a land connection across Antarctica, the Early Cretaceous terrestrial vertebrates of Australia seem to have greater affinity with those of Asia and North America. This may reflect the Noah's Ark phenomenon, where micro-continents broke off the northern edges of Australia–New Guinea and drifted north to become incorporated into Asia. In doing so, these "Noah's Arks" may have carried elements of the Australian biota into Asia and thus the northern hemisphere.

For the Future

The Dinosaurs of Darkness project is not yet finished. Digging will continue at the Flat Rocks locality for as long as we can continue work there. For many years, the search for Australian Cretaceous birds and mammals rode on a dinosaur's back. That is, we could justify to ourselves the effort to continue the search for Cretaceous birds and mammals because we were continuously finding polar dinosaur specimens in the process and learning all sorts of fascinating new things about them. Now the reverse is true at Flat Rocks. The learning curve about the dinosaurs at that site seems to have flattened out, and it is our judgment that the effort necessary to get further specimens of only dinosaurs there would *probably* not be rewarded with enough scientifically significant results to justify the additional work. We hope that we are wrong in that judgment, and perhaps something as worthwhile as a partial dinosaur skeleton will eventually turn up there. But we shall keep digging up dinosaurs there now because the primary objective has shifted to the recovery of the mammals and birds we know to be there. We are right at the bottom of the learning curve for them. More dinosaurs will be a by-product of this effort to find the answers to questions such as whether the upper molars of *Ausktribosphenos nyktos* support or refute the hypothesis that it is a placental mammal, and if so, whether they are particularly closely related to hedgehogs. The discovery of more dinosaur fossils at Flat Rocks could well result, for example, in a statistically large enough sample of femora to at last determine just how many different kinds of hypsilophodontids there were in southeastern Australia during the Early Cretaceous.

Questions raised by these Victorian polar dinosaurs have caused us to become involved in a number of related projects elsewhere. Within the Mesozoic of Australia almost all the dinosaur skeletal material is between 95 and 115 million years old. That leaves about 80 percent of the dinosaur record represented by only a single Middle Jurassic partial dinosaur skeleton and the tracks of Triassic and Jurassic dinosaurs. That one skeleton is the lure that will draw us in the future to search for earlier dinosaurs and mammals in southeastern Queensland, where it was found. We know that at that site fossil bone has been preserved and can still be collected.

As we examine a paleogeographic map of the Early Cretaceous Earth, it seems that it was possible for dinosaurs to have passed between South America and Australia via Antarctica. But did they actually do it? As yet, if anything, the dinosaurs of that age in South America seem to have been isolated from those of Australia. But then again, the Early Cretaceous record of South American dinosaurs is quite meager.

At the invitation of Director Rubén Cúneo of the Museo Paleontológico Egidio Feruglio in Trelew, Chubut Province, Argentina, we have been searching with his staff for Early Cretaceous dinosaurs in Chubut since 1994. Thus far, we have found a few sites worthy of further attention but precious few fossils (at least by Argentinian standards, certainly not Australian ones). The most important advance made has been to identify which rock units and areas are worthy of intense prospecting. Just as the Victorian dinosaur assemblage is biased toward small bones, the Early Cretaceous dinosaurs of Chubut are biased toward large ones. That we have found no particular evidence for overlap as yet may simply be because of a difference in the mechanical factors that sorted the bones of dead animals in the two regions rather than a fundamental difference between the animals that once lived in those areas or, more likely, because of the small samples we have. At this time we simply lack the evidence to decide.

The comparative method is an obvious approach to understanding the paleobiology of dinosaurs that lived at high latitudes. The richest known polar dinosaur assemblage is that from the Colville River on Alaska's North Slope, specifically, the Liscomb Bonebed.[70] It is an incredible accumulation of hadrosaur, or duckbill, dinosaur bones that extends for about 100 meters along the banks of the river. It appears to be the result of a mass drowning of a herd of these animals, perhaps as they were crossing a river. Besides hadrosaurs, the North Slope has yielded a suite of other dinosaurs, particularly ceratopsians. To date, large theropods are only known there on the basis of isolated teeth that have been allocated to small *Albertosaurus*.

While we were there in 1989, we met up once again with Roland Gangloff of the University of Alaska, whom we had known when we were all students at the University of California–Berkeley. Together we have formulated a plan based on the tunneling experience we have gained at Dinosaur Cove.

One of the problems with working the outcrops along the Colville River is that they are formed of permafrost. Under the summer sun, the steep permafrost slopes melt and collapse. Also, the repeated freezing and thawing of the bone in the zone where melting occurs causes it to break up. We would like to try to tunnel into the permafrost during the spring when the permafrost is still frozen solid. The tunnel would be constructed above the layer where fossil bone is known to occur. An entrance structure would be constructed like the door to a freezer. Once the tunnel was fully constructed, the door would be closed until the following summer. At that time, a field party would return to excavate the floor. The air temperature inside the tunnel would be –3°C (27°F) summer and winter. There would be no melting of the permafrost inside with the door shut nor would there be slush or rain or wind

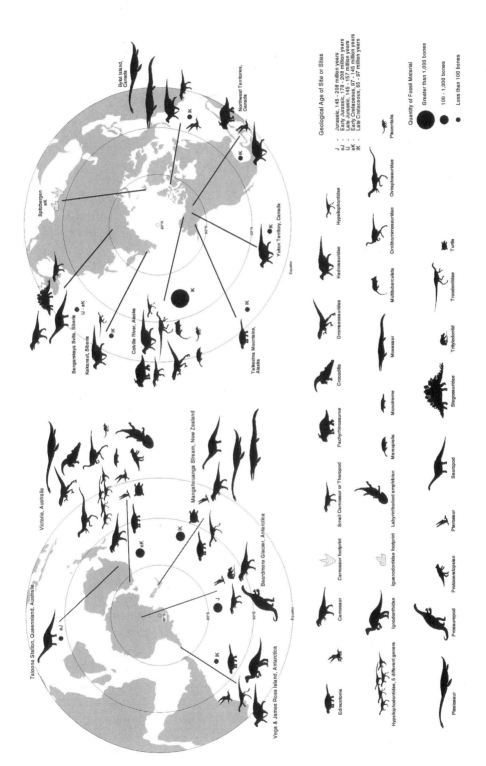

Figure 68. Jurassic and Cretaceous polar latitude tetrapod localities. The base map is drawn for the Early Cretaceous. For this reason, the Middle Jurassic sauropod shown from southeastern Queensland, Australia, appears to be outside the polar region whereas, in fact, when that animal was alive, it lived virtually on the South Pole. A. Southern hemisphere. B. Northern hemisphere.

Figure 69. The left bank of the Colville River, Alaska. Formed of permafrost, it melts under the summer sun and quickly forms talus slopes. All the talus in this photograph accumulated in the two or three months between when the ice on the river broke up in late spring or early summer and when the photograph was taken in late summer.

to hamper the work. Finally, the bone would never have been in the freeze-thaw zone. Therefore, it would be intact instead of cracked by that process.

We know this because in 1994, Roland and Tom, together with two experienced Arctic miners, Mike Roberts and Earl Voytilla of Anchorage, Alaska, visited the Liscomb Bonebed site with funds provided by the Dinosaur Society and went over the proposed plan in detail. With a few minor changes, Mike and Earl pronounced the plan quite feasible. Except for the fact that our objective was dinosaur bones instead of gold, what we proposed to do was "off the shelf." That is, we did not have to develop any new techniques but rather simply borrow those long used by gold miners in the Arctic who were after alluvial gold in permafrost. When they returned to Anchorage, Roland and Tom were invited into a gold mine that Mike and Earl operate there that had been cut into permafrost. The conditions inside were certainly far more suitable for excavating dinosaurs than those experienced just a few days before on the banks of the Colville River!

Figure 70. The entrance to the gold mine of Mike Roberts and Earl Voytilla, near Anchorage, Alaska. The tunnels of this mine are cut into permafrost. This type of entrance with an insulated door keeps the temperature inside a uniform −3°C (27°F) all year round.

Figure 71. Inside the gold mine of Mike Roberts and Earl Voytilla.

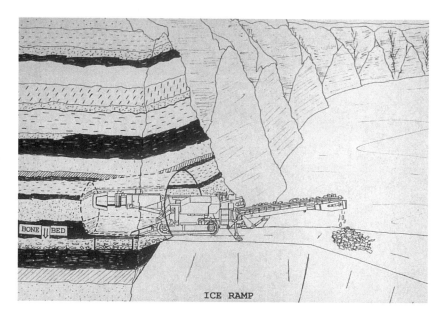

Figure 72. A road header as it might be used to excavate a tunnel above the known dinosaur-bearing permafrost on the Colville River, Alaska. In the spring, the equipment could be driven over the 2-meter- (6-foot-)thick ice on the river right to the locality.

Working in permafrost is not like digging in Mongolia or Montana. Techniques need to be developed that take advantage of it rather than trying to use methods developed in temperate latitudes that are inappropriate in the polar regions. What Roland and we hope to do is to eventually find the funding to carry out a trial dig to test the feasibility of this approach. If it works, the 200 km (125 miles) of outcrop on the left bank of the Colville River that spans the entire Late Cretaceous will be much more readily investigated scientifically than at present. From it could come a sequence of polar dinosaurs which could show us their evolution through time in a way that no other place we now know of has the potential to do. That would be a very suitable yardstick for comparison with the Victorian polar dinosaur assemblages.

Because of our general interest in polar dinosaurs, we hope to work with Roland and with Bill Hammer, who has collected Early Jurassic dinosaurs under most difficult conditions high in the mountains of central Antarctica, to create a traveling exhibition about these animals that would go on the road in 2002.[71] That would enable people not only to read about them but to see them as well. It will also provide some of the funds needed for future work on the Dinosaurs of Darkness, which still need much light thrown on them.

13

Afterthoughts

The Burning Bridges behind Us

Just as Eliza, the runaway slave in *Uncle Tom's Cabin,* could not follow the same series of ever-shifting ice blocks twice across the Ohio River, we are well aware that if the Dinosaurs of Darkness project were beginning today instead of in 1978, it would be done quite differently. That things have changed significantly in that time goes without saying. This raises the question, have the changes that have occurred in the scientific community specifically and society in general since then made it on the whole easier or more difficult to do innovative research? This question is not an idle one, because it gets to the heart of a matter of broader concern. Five interrelated issues are relevant to this question: safety standards, the cultural commitment of society, child care, the remuneration of scientists, and trust.

1. *Safety.* Could the Dinosaurs of Darkness project be started today? We often wonder what would have happened if the work at Dinosaur Cove had not been done when it was. Would we even be able to contemplate doing the work today? Dinosaur Cove was at the bottom of a 90-meter (300-foot) cliff right at sea level on the rugged coast of southern Australia—a sea wide open to the great southerly winds that roar in from the south and southwest, sometimes straight off Antarctica. It is not a forgiving place. Many people have lost their lives in great seafaring tragedies. Many people have drowned along this coast, some swept off coastal rock platforms by unexpected waves, some because they simply were not careful in their assessment of the unforgiving nature of the ocean.

Today in most universities and public institutions across Australia, and we suspect that this may well be the case elsewhere in the world, there is a great move to make sure that all staff members carry out their duties in a completely safe fashion. Usually there are safety officers and rulebooks about how one is to carry out fieldwork. There is nothing intrinsically wrong with this—that is, if it is not carried too far. Within the environment in which we now work, we are not entirely convinced that the ever-increasing rules would allow us to carry out the soft-rock mining that we did at Dinosaur Cove with the large crews of volunteers. Maybe yes, maybe no—there is really no way of testing this. The fact is that we carried out our excavations there for over a decade, using explosives and working in an extremely taxing environment, with no serious injury. We used common sense, we were careful, and, most important, we cared about the personal well-being of each and every person that helped us.

Pat O'Neill is to be credited for his constant solicitude of us and this site. He was a "safety officer" par excellence whose repeated personal visits to the site in order to evaluate both us and the procedures being used is not the approach to this job that most people who are doing it today take. Rather, because of their lack of qualification in this specific area and the tremendous workloads they carry, they are rulebook oriented and office bound.

We were gratified to learn, after the fact, that the Research Committee of the National Geographic Society had once evaluated this aspect of our project. Based on what they knew personally about both of us and how the work was proceeding, they were confident enough of our abilities in this area that they fully supported us, both monetarily and morally, despite the dangers involved.

To us, this would be an excellent approach for the safety officers of institutions to emulate, that is, to evaluate the researchers in charge of projects. The safety officers should be sure that the researchers know what the commonsense rules are and know the legal issues. On that basis, safety officers should advise if there are aspects of the project that need some attention. Asking staff to attend endless safety sessions, gain qualification in numerous courses, and so on, just to run a field crew may be excessive and really may not sort the sheep from the goats in any case. We can imagine many people who, having completed all these requirements, might find themselves at sea when faced with evaluating a real situation or dealing with a genuine crisis. In order to be able to evaluate the situation, our approach would mean that the safety officers would have to leave their offices, go into the field on occasion to observe what is going on, and make it their responsibility to know their researchers—a cooperative sort of arrangement. Fortunately at Monash, where Pat works, this is the case.

You can only put so much armor on a battleship before it sinks. A compromise in design has to be reached between the various objectives in constructing such a ship or you will have a battleship at the bottom of the ocean that cannot carry out the mission it was built for. Our hope is that the current state of rules and regulations that are being applied across the board to fieldwork will not become so detailed and ponderous that they actually stop many innovative projects that could be safely carried out and that have so much to offer in the future for the advancement of science and the society that supports it. Interpretation of the rules relative to individual projects appears to us to be an area that requires sympathetic and in-depth thought by both safety advisors and researchers. The Pat O'Neill model is a good start.

2. *Cultural Commitment.* On a visit to Russia in 1994, one of the things that struck us both was the commitment of the people we met to their culture. Despite the hard economic times they were going through, these people felt it was important to preserve and protect the things that made Russia Russian. Restoration of churches, whose construction had stopped during World War I, was going on everywhere. This was not done by people who had carried out market surveys to find out who might pay to come and see these buildings, or who had made a profession of consulting with people to identify stakeholders. It was done by people who did it because they knew it was a good thing to do, by groups in the towns who knew it was the right thing for their towns.

When we first lived in Australia, things were done with much the same attitude. People seemed to know what should be done and did it. If there was a dollar in it, well and good. But if not, it was done anyway. Unfortunately, in the past twenty years, the concepts of economic rationalism have found exceptionally fertile ground on this continent. With the one-dimensional view that money is the measure of all things, the current thinking is all too often that if no one will pay for something, obviously it is not worth doing. This sort of thinking leads to trying to sell what will bring the greatest immediate profit, not in recognizing what is good for society in the long term. To cherish and put resources into what is good for society in the long run, even if it is not a salable commodity in the short run (or for that matter a salable commodity at all!), requires being able to identify those things that possess this attribute and protecting them. This, in turn, means that there must be people who have this exceptional foresight and who have the wherewithal to do something with their ability. Those people must be encouraged and nurtured if society is to have any real values and a future that is not a bleak, sterile existence based exclusively on dollars, yen, or rubles. In the long run, it is not certain that a society with no other values than those that can be measured with dollars can survive.

3. *Child Care.* Pat has often commented about opening her mail or her e-mail and finding much information about support for women to give them equal opportunities or to help them develop leadership skills.

She has sat through a number of seminars and participated in forums on this subject, and has noticed an issue that is very frequently not discussed, or if discussed, only briefly—child care.

The single most useful act that would have helped both of us immensely in the carrying out of our research and field lives would have been, and still is, superb and continually available help with our children. Pat suspects that this is an issue that affects a large number of professional women in science (and elsewhere), and men as well who find themselves the primary caregiver for children in a family.

Pat's parents, Mary Lee and Reid Macdonald, helped us out on several occasions with child care, making it possible, for instance, for both of us to visit the North Slope of Alaska at the same time. They even traveled to Australia once for the sole purpose of helping out with a dig. However, because they live in South Dakota, there have been many times we have had to hire other help with our personal resources if fieldwork was to be done by the two of us together. Many of those people who assisted in this way have become part of our family. We owe a great deal to these people, especially to Pat's graduate students (Corrie Williams, in particular, who has persevered for many years and has become an honorary aunt as far as we are concerned). People in the areas where we have lived have also helped. Faye Ovenden, who lived near us in Emerald, accompanied us several times to Central Australia specifically to care for Leaellyn, beginning when she was only one year old. Also, people on our field crews have taken our kids on board as part of their lives in the field, but we have always tried to be sure that this does not become an onerous task for the crew.

Especially when we were working in more remote areas or at times when Pat was cooking full time, managing the camp, and trying to also do some fieldwork—when Dinosaur Cove crews had upwards of thirty or more members—it was damned difficult to do all that and have two small children at heel and it led to extreme exhaustion. It would have been wonderful to have had some monetary help to hire a child care helper.

If this is what society demands of its scientists with parental responsibilities, then so be it, but we feel it would be a reasonable step to allow research grant applicants an item for child care—including "in-the-field child care"—as a legitimate budget item. It has far too long been a taboo. It would have been quite possible for Pat to acquire funding for a professional development course in leadership. But to get child care funds to allow her to carry out her work in remote Patagonia would have been impossible. Some years it was impossible to manage kids and remote fieldwork, so she just stayed home. In other years, she took the kids in the field, at times with one on her back and the other trailing along up and down the hills with hand in Mum's. We both know what Sacajawea must have felt like as she walked from the central United States to the west coast with a small child on her back while she served as interpreter to the Lewis and Clark expedition.

The Demos think tank in Britain, involving the home secretary of Tony Blair's government, Jack Straw, has published a series of recommendations about policies that should be implemented allowing flexibility of schedules to support professional people in raising children as responsible parents in society and still making major professional contributions.[72] At least one government is taking a longer and broader view of these issues, and we strongly urge a closer look at their detailed recommendations about parental leave and flexibility of schedules. These are areas where basic policy changes need to be made by government and employers—changes which recognize that children and professions can co-exist and that both have intrinsic value. Both need attention and time devoted to them without sacrificing one for the other. Governments and employers will only make such changes when society in general comes to this view and strongly pressures for change.

4. *Remuneration of Scientists.* Scientists should not be paid too much, because then people whose primary motivation is money will enter the profession for the wrong reason. However, those who fund the scientific enterprise do not have to worry about that difficulty any too soon! To have the science done that is necessary for a society not only to survive but to flourish, those who are most needed for the enterprise are often the people who will do it anyway under all but the most adverse conditions. A more adequate reward than what is currently available, particularly for young scientists, will markedly increase their capacity to accomplish their tasks. We fear that with the lack of current tenure-track jobs for young scientists, fewer and fewer will enter that profession. So often now, a science Ph.D. faces years and years of two- or three-year post-doctoral fellowships and no guarantee of ever getting a permanent job. So many bright minds turn to computers or business consultancies of various sorts. Is this where society wants to go: travel, entertainment, and business and fewer and fewer scientists who can tackle the fundamental, long-term problems that must be solved in order for us to survive and flourish? We think not. And if not, there need to be more secure jobs at the end of the Ph.D. in science with pay comparable to jobs in these other fields.

5. *Need for Trust.* The heavily bureaucratized way of dispensing most research funds today is not working very well. A greater and greater percentage of the total amount spent on scientific research goes to administering what funds are left, at least in the Australian system. The publicly stated motivation behind this bureaucratization is to make the system both fair and accountable. However, those who wish to cheat the system will. Such people can be quite clever and readily find ways

to do so. In the meantime, those who are more interested in getting on with the job are stymied by an ever-enlarging, inextricable tangle of bureaucratic red tape. A system with more trust would not be perfect and would undoubtedly be abused. However, a description of democracy attributed to Winston Churchill might be applied here. "Democracy is the worst form of government in all the world, . . . except for all the others." Similarly, a system of support for science based on a large element of trust might be the worst system in all the world, except for all the others. Perhaps this is where private funding agencies have their greatest role to play in the future. Because they are not accountable to the general public, they can base their system of dispensing funds on trust in a way that is not politically possible for government-funded agencies. This means that really innovative science, in at least the pioneering stages, will not often be done under government auspices in the future. As we read about how the transnational corporations are displacing nation-states in political power on the world stage, is it the lack of widespread accountability on the part of transnational corporations that gives them their edge? Those who wish to preserve democracy as more than a shell should perhaps consider this problem.

The Demos group has also considered this topic in detail and has concluded that somewhere trust must come into the story.[73] In our own experience, the problem of the audit has increased our workload a great deal over the past decade. We still do not understand when we write and are successful in bringing in a grant, especially one that requires only the most general of accounting by the fund-granting agency, why it is necessary to keep incredibly detailed records for the institutions who administer our grants of absolutely everything we might buy on a field expedition. Why a reasonable per diem rate for a trip cannot be agreed upon by the institutions where we work boggles our minds. On most days we would not even reach the standard per diem, while on those days that we might exceed it, we would never think of asking for more, since the grants are of a limited nature anyway. Why some institutions allow administrators to use a per diem rate while they do not allow scientific staff to do so also puzzles us.

Fortunately, at Pat's institution, this trust does exist at high levels of administration. But at middle-level management there is often a real lack of understanding of the manner in which field research is carried out. The audit system must have been solely designed for those traveling and working in urban environments with a sophisticated financial infrastructure in place. Trying to apply such a system to the isolated hills of Patagonia or central Australia simply does not work efficiently at all. The end result is the dissolution of trust of the scientists by some people in middle management, the waste of a great deal of time in the auditing process, and the gnashing of teeth and tearing of hair of those trying to do original research on a minimal budget while being overwhelmed with red tape and being treated as if they are trying to cheat the system.

Another great unnecessary drain on limited resources of the modern working scientist is the need to generate an excessive number of internal reports which serve little or no discernable purpose. Often the reports requested are permutations of information already presented. The administrators are simply unable or unwilling to go to the already submitted reports and dig out the information they require. Why do so? It is so much easier to ask for another report because it is "free"; administrators have the ability to require such documents of others on demand without having to provide the funds needed to carry out the task. We offer a modest suggestion. One solution would be to make such internal reports no longer "free." Just as modern management practice calls for payments between different components of organizations for services rendered by one division for another, divisions calling for such reports from other divisions would have to pay for them. This way, there would be a strong motive to economize in this area on the part of people seeking such information just as there now is so frequently in the use of stationery, telephones, and postage. In this case, if a report was needed and funds were not available, the persons seeking the report would either have to find out the information in some other way or obtain a budget allocation to reimburse those from whom the report was sought.

We can attest to just how demoralizing this has been over the last decade. It has sometimes brought us to the brink of throwing in the towel. As we remember some of the times when things were at their worst, we are so heartened that we did not just walk away from the whole project. But we must say, however, that these sorts of frustrations came close to stopping us—more than not having funds, more than the difficulties of not finding things for so long—the red tape and the total lack of trust of the middle management administrators nearly killed the whole project.

These were the times when the solid and unwavering trust given us by the National Geographic Society, the vast numbers of volunteers from Australia and around the world, and our faithful lab staff sustained us, as did Pat's deputy vice chancellor, Peter Darvall, and her vice chancellors, Mal Logan and David Robinson. These were the people who truly kept the scientific program alive. Trust is very important and not to be underestimated. Mistrust can fundamentally undermine the human spirit.

The Role of Sponsors

Be it the painting of the ceiling of the Sistine Chapel or the discovery of the neutron, advances in the arts and sciences do not take place in a vacuum. To do either of those things required that someone devote inordinate time and energy to something not directly related to staying alive. This meant that they or someone else had to have the means of providing the creators or discoverers with the wherewithal to stay alive while engaged in their activity and provide them with the means to do the activity itself. The search for polar dinosaurs in southeastern Australia was one such creative act.

The initial phase of the search for polar dinosaurs was kick-started by the Australian Army. On his first day on the job at the National Museum of Victoria, Tom had a meeting with a warrant officer from a unit of the Royal Australian Electrical and Mechanical Engineers (RAEME). The purpose was to come up with a plan for a combined trip of army and museum personnel. The objective of the trip was up to Tom. He took the opportunity to return to sites we had been to a few years earlier on our first trip to Australia. On this trip, fossils were collected and additional sites were found. It laid the groundwork for the continuation of the prospecting effort that many years later would lead to the discovery of Dinosaur Cove and, later, Flat Rocks.

The army's objective was to get its personnel doing something outside of their routine. This was both to challenge their professional abilities by putting them in unusual circumstances and to encourage them to remain in the army. As sponsors, the army had their agenda, but at the same time they provided support for fieldwork that otherwise would not have been available so quickly to a new curator. We eventually recognized the army's assistance in this way when we named a Miocene mammal from one of the sites visited on that first trip: *Raemeotherium.* This was done after the event and had not been requested by the army. It was our feeling that help should be recognized, and this was a meaningful way to say thank you. Provided civilization stands, that generic name will still be around in a century's time, long after the RAEME-sponsored trip of 1974 would likely have otherwise been totally forgotten.

Four more times over the next twenty years, the Australian Army helped out in a similar fashion. The most difficult thing we ever did with the army was to explore dry lakes in the Canning Basin of northwestern Western Australia. Geological reconstructions of the area indicated that the vast sinuous dry lakes that are found in that area today were broad flowing rivers during the Cretaceous and early Cenozoic. All were mapped as having Pleistocene surfaces. But we hoped that perhaps those maps might just be wrong in some places. Older clays are known to exist at depth beneath the floors of those dry lakes. Because much of the mapping was done by analysis of aerial photographs, it was just conceivable that somewhere older clays had been uplifted on the floors of some of those lakes and simply overlooked because no one had ever walked over them specifically looking for fossils. R. A.

Stirton and R. H. Tedford had found late Cenozoic mammals in clays exposed on the margins of dry lake beds in South Australia, so it did not seem out of the question that similar geological circumstances might occur on the margins of those dry lakes in Western Australia. A fortnight's search, however, failed to reveal a single example. This time hard work in difficult terrain yielded nothing.

It was not for lack of trying that this was the outcome. Major Russell, the commanding officer of the Pilbara Regiment, did everything he could to get us to where we wanted to go. The first day of the trip began with us rising at 5:00 A.M., packing up, and getting under way by 5:30, when it was still pitch dark. With two breaks during the day for refueling (we ate as the twenty-two-vehicle convoy roared eastward), we drove 1,500 km (930 miles), of which 900 km (560 miles) was on dirt tracks, before we stopped at 10:00 P.M. that night. From that point, Well 33 on the Canning Stock Route, we drove another 300 km (185 miles) to the area where the prospecting began. In order to keep the field party of more than forty people going, Major Russell had a number of trucks driving around the clock, fighting their way across the sand dunes, to bring rather brackish water from Well 33 to the advance camp. Other officers, such as Captain McMillan and seasoned warrant officers like Scruffy, got us through other sand dunes and over the muddy areas on the lake floor to wherever we wished to go. They also made it possible for us to do things we had not anticipated doing when the trip was originally planned. In order to take sediment samples at depths of 1–2 meters (1–2 yards), they and their men set to with a will and dug holes in the sticky mud with shovels—not coring rigs!

Disappointing though the outcome was for us, the time spent was worthwhile—we could write off a potential area that had long beckoned us. For the army, this trip achieved what they were after, because their men had successfully carried out a complex operation in difficult country. Thus, as sponsors, they had achieved their goals as well.

In all cases, the assistance that the Australian Army provided was full logistical support and personnel. All the museum scientists had to do was show up and say what they wanted done and, within reason, it was done.

Over the years, many other organizations provided gratis services and equipment that was in their normal product line. For example, Safeway Australia willingly gave us stock with damaged packaging, even though the stock itself was still good, to use at Dinosaur Cove. Each week we went to their warehouse and sorted through what was available. In this way we kept the crew well fed. Other organizations that helped in this way were Shell and Mobil with fuel, Orica (then ICI Australia) with explosives, and Ingersoll Rand with split sets. This type of support in kind was vital to the work. But so, too, was cash. The most important difference between in-kind support and cash is flexibility.

When one has more tires than needed and lacks food, it is not always a straightforward matter to barter one for the other.

In the initial phase when we were groping for a way to locate Mesozoic or early Cenozoic terrestrial vertebrates, we applied for small grants from both the Australian Research Grants Committee and the Committee for Research and Exploration of the National Geographic Society. The grants had two intertwined objectives. The first was to search for the desired older sites. The second was to visit areas where mid- to late-Cenozoic terrestrial vertebrates had been found by Stirton and his students and later on the RAEME-sponsored trip. By doing the latter, we hoped that just as Stirton had done, we might find older localities where the younger vertebrates occurred. There was another reason as well. Although the broad outline of mammalian and avian evolution during the past 20 million years on the Australian continent was known by the time Stirton died, there was still much to be learned. By collecting in those younger deposits where, in the short term, fossils were more likely to be found, we had something to work on while the search continued for the desired older specimens. It is much easier to keep trying to find elusive fossils if one knows that there are others already collected awaiting study and analysis, even if they are not exactly what we wished to have, given the choice. It also meant that we still had access to funding, for most agencies supporting scientific research insist on visible, immediate results in the form of professional publications.

In the exploration stage, the support we sought was modest, for two reasons. First, the surveys involved small numbers of people in areas no more than a few days' drive away. Second, at that time Tom had technical staff at the Museum of Victoria to assist him. The staff was critical in readying the fossils collected for scientific study. They did this by carefully removing the rock from around the fossils once they had reached the museum. They also assisted in illustrating the fossils for eventual publication.

Over the years the personnel provided to Tom by the Museum of Victoria to assist in this work gradually dwindled away to nothing, and alternative help had to be sought. Preparators and assistant curators were not replaced and exhibitions, instead of being seen as equal in importance to scientific research, became seen by the senior management as the raison d'être for the museum. The senior management in this period came to their positions with no previous museum experience.

This gap was filled for many years by the Australian Research Grants Committee. However, a gap of four years in such support nearly sank the project. Only by Pat's ability to piece together small amounts of money from a variety of sources, such as book royalties, public lectures, and by taking on additional work, was this critical function maintained. In addition, we sold our home to plug the financial holes in both research and development of a center for science education that was poorly funded at the state and federal levels. That was the end of our personal assets, except for a 1976 Volvo and our books. The outcome, however, was encouraging—the research work continued and the Monash Science Centre for public science communication survived despite the times.

What's in a Name?

Maps of Australia drawn in the 1830s showed an accurate outline of the coast and the inland areas around the major cities. The details become sketchier farther inland, until there is nothing but white paper in the center. To the Europeans then living in Australia, that expanse of white paper was a great unknown, constantly motivating them to send exploration parties into what came to be called "The Ghastly Blank."

Because our research program hoped to learn about the origin and early history of Australia's unique mammals and birds—which were blanks in scientific knowledge at the time—our proposals for assistance to the National Geographic Society were given the name "The Ghastly Blank."

The members of the Committee for Research and Exploration of the National Geographic Society were fascinated by the prospect of what we might find as outlined in these "Ghastly Blank" proposals. They backed our efforts almost continuously for more than twenty years. Even during those years when little was discovered, despite the incredible amount of work being done by large crews of volunteers, the National Geographic Society not only continued to fund the work but personally encouraged us as well. In particular, Frank Whitmore, a longtime member of the committee, and Barry Bishop, who was chairman, never lost hope that eventually the mammals and birds which had been our original objective would turn up. By the time such fossils were found, we had long since resigned ourselves to being content with polar dinosaurs. They are certainly quite intriguing in their own right, but they were not the original objective, which we never forgot, despite the frustration of two decades of searching and not finding.

One of the real paradoxes in writing a grant proposal is that the people who may fund the work naturally want to know what results you anticipate achieving. One can only guess and come up with a plausible outcome. But if a grantee could really outline the results in detail in advance, it would not be research and there would be no point in carrying out the work. It is because the unexpected happens that scientific research is worth doing. When Darwin joined the *H.M.S. Beagle* as a naturalist, he did not justify doing so by saying that he was going to formulate the theory of organic evolution!

Fortunately, experienced people in the granting agencies realize this. Even when we changed the emphasis of our research program, we never had to apologize to the National Geographic Society, the Australian Research Grants Committee, or the Monash Small Grants Committee that while we had started out to discover early Cenozoic and Mesozoic Australian birds and mammals, what we actually did was to become involved primarily with polar dinosaurs. We certainly did not want to walk away from a topic so intriguing and none of the project's sponsors expected us to. It was only because we persisted with the alternative project that we eventually achieved our original goal as well. Scientists seize unexpected fruitful research opportunities all the time. But the terms of reference of the granting agencies do not always encourage such a shift. Had the change in the research program, for example, led us not to another area of paleontology but into meteorology, for example, the funding bodies to be approached would have been totally different. It is unlikely that they would have been sympathetic to proposals that we do research outside our field, no matter how exciting the prospects were.

Besides the physical wherewithal that makes research possible, there is another factor that is equally important: time. For almost all of the first two decades that Tom was at the Museum of Victoria, he did not have to justify the project to museum management. They assumed that what he did as a vertebrate paleontologist was up to him and that he would as a matter of course do what a vertebrate paleontologist should do. There was a high element of *trust* in this arrangement. Not having to be accountable for every detail during the formative period of this project was essential for the simple reason that the way ahead was not clear. The book *In Search of Excellence*[74] surveys what makes some American corporations highly successful. The authors found that the research programs of the successful companies were typically championed by a powerful person in senior management. That person's role in successful research programs was to keep loose tabs on what was going on and provide the researchers with resource support outside of the normal channels within the organization. In essence, this is what the arrangement was at the Museum of Victoria during the early stages of the project, as was the case for Pat at Monash University. Her champion was Professor Jim Warren, the chairman of the Zoology Department, and Professor Bruce Hobbs in the Earth Sciences Department—both of whom were her bosses.

It is a common complaint that the burgeoning amount of scientific publications that a scientist must keep up with is due at least in part to the mandate to "publish or perish." That phrase accurately encapsulates the reality that if a research scientist does not publish frequently in recognized scientific journals, their career may well come to an end or at least the scientist will certainly not be blessed with funds to carry on research. This results in the writing of many scientific papers for which there is little justification. Many in this genre simply repeat in different words a paper written earlier by the same author. Fortunately, this never was a serious problem for us. As late as 1990, for example, Tom did not publish a single paper because he had nothing to say that would have been a novel contribution, the essence of a worthwhile scientific publication. We were busy trying to do the work so that contributions of substance could be made in the future. Tom did not divert his time to writing an unnecessary paper, a "potboiler," merely in order to show that he was still an active scientist. That this was possible without condemnation is a reflection of the enlightened viewpoint of the sponsors and the attitude of the employers we had at the time.

As the project has matured and the goals have become clearer, detailed plans can now be made well in advance. It is fortunate that this phase was reached when it was, because the management style of most institutions has changed. Now there is a major emphasis on project management and formal accountability. Now much time and effort are devoted to writing detailed research plans, yearly evaluations of proposed goals, and so forth. In short, scientists today often find themselves caught up in corporate plans that might be appropriate to those working in very structured repetitive environments where detailed accountability for meals, hotel rooms, and daily expenses is a straightforward matter. But field research scientists are not bureaucrats carrying out standardized structured jobs. They are often working under extremely unpredictable conditions, the very unpredictability associated with scientific advances. They are often doing this work in circumstances that cannot possibly be anticipated while trying to be responsible parents, to nurture students, and to come up with the novel ideas that are the hallmark of the best science, in addition to doing the hard work of documenting and testing those ideas. If the leaders of our society want to kill such creativity, then all they need to do is to continue along the trail of red tape that has increasingly been laid over the past decade. We think that will do in people like us quite nicely. We just wonder where the new ideas and imaginative young scientists will come from if this trend persists.

Had that system been in place in the 1970s and early 1980s when the way ahead was quite unclear, it is difficult to imagine how the critical formative stages of the project could have been carried out.

The Media

The popularization of science is certainly not new. The nineteenth century saw figures like Thomas Henry Huxley, Darwin's Bulldog, devote much effort to presenting the results of the latest scientific research to the lay public.

What is certainly new are the various kinds of media that are now available. Radio and television documentaries and Internet websites are quite different ways of getting a message across to the general public than are the print media. And within those various electronic means of communication, a number of different procedures are used.

Although the equipment used is identical, it is, for example, quite a different matter to give an interview to a television news team than it is to participate in the making of a scientific documentary. The first may go to air in a few hours, the second may not go to air for a year. Time is of the essence for a television news team. As a consequence, the results are often mixed. Sometimes the information is accurately conveyed and one is quite relieved. At other times . . .

Makers of television documentaries are not trying to get something out for the evening news. However, they are always working within a budget and invariably it is not enough to do all the things that the producer would like. To go as far as possible, they, too, try to get the work done in the shortest possible time. This often results in a lower-quality program than would be the case if the initial plans were more modest for a given amount of funding. Such a plan would mean that a significant amount of surplus time was built into a production budget as a matter of course to adequately cover the inevitable, unexpected changes and opportunities. Being able to take advantage of the unexpected is what can make a television program great. Beyond this factor, there is a critical division among television documentary makers. There are those who want to get the information correct at all costs. They will settle for less as long as it is correct. These documentary makers arrive on site and discuss what is going on for some time in order to fully understand the situation before they decide on the sequence of shots they want to make. The second category consists of those who arrive on site with a fixed idea of what they want. Sometimes what they want is based on a very superficial understanding of the topic. They will often bend the facts beyond the breaking point in order to fit their preconceptions into the finished product. Some of these people can be gently educated in the process of making the documentary, and the result is a better film.

Of necessity, all media people covering a scientific story are doing so because it is their profession. At the end of the day, they are there to make a living. However, the relationship is not all one way. There would be little point in doing what we do if other people did not learn about it. While those other people include our scientific colleagues, they are not the only audience we hope to reach. People generally have a natural curiosity about who they are and what their place is in the natural world. Science addresses that concern. As we, too, are part of the human race, and also because we have been supported in our efforts by elements of human society, we feel an obligation to impart what we have learned to people generally so that all may benefit. Many of those in the media with whom we have worked are motivated by the same feelings.

The most encouraging experience that we have had with the media came from a long association that our project had with the Australian Broadcasting Corporation (ABC) science series, *Quantum*. In the early 1980s we had the profound good luck to be approached by this group to do an in-depth story on the work at Dinosaur Cove. A young presenter, Karina Kelly, came down the long rope into the cove to begin her interviews, and we instantly fell in love with her—that is our young son Tim did—he often remarked that he loved Karina. Karina's father was a physicist and her husband, David Mills, was too. Karina fundamentally understood what science was about and was an extremely intelligent science communicator. Karina and *Quantum* possessed another attribute which we immediately appreciated—she came to film us and interview us doing what we do without being overly intrusive or trying to script us. She quickly sorted out who were the people who felt comfortable in front of the camera and immediately put us at ease. She and the crew stayed awhile, not rushing in at the last moment and rushing off—they took the time to do this.

We all admired Karina and the *Quantum* crew and truly took them on as friends. This bond still exists today with Karina. *Quantum* did another interview some eight years later and an in-depth interview with Pat about what it was like to be a scientist. Karina also came along on the trip to the Cape York Peninsula and single-handedly directed, scripted, narrated, and produced a documentary about it. As a result, the programs *Quantum* produced were of very high quality and really captured the excitement and the essence of the excavations at Dinosaur Cove. We have always appreciated the media interest because it is a way to let the public know what we were doing and let Australians and the rest of the word know the significance of the discoveries along Australia's southeast coast. Even more important, the audience gained an idea of how science is actually done.

Dinosaurs in Cyberspace

http://www.earth.monash.edu.au/dinodream/—a place in cyberspace that tells about the latest discoveries at our Inverloch project, named by Lesley "Dinosaur Dreaming" after the Aboriginal dreamtime. When we started digging at Dinosaur Cove, the whole concept of cyberspace, of Web sites, was in its infancy, but that was not so as we began intensive work to the southeast of Melbourne at Inverloch. Longtime dino-phile Marion Anderson (one of our volunteers) and her husband Phil are both very facile with the mechanics of setting up Web sites. They suggested early in the excavations at Inverloch (in which they were both also fundamentally involved) that we should set up a Web site that would be updated regularly. Accompanying this was another idea, that of setting up a Friends of Dinosaur Dreaming organization, to be made up of people who want to participate in the dig or people who just want to know about the dig and offer some annual financial assistance to help keep the work going. Marion and Phil pursue the information ex-plosion on the Web for us and keep us informed about other sites and data available on the Web—a time-consuming, but fun, endeavor. Mar-ion is a particularly unique asset because she also coordinates the large first-year geology course at Monash University, where students defi-nitely hear about dinosaurs in Victoria—and from this group come many new volunteers. Phil works as a cyberspacer and his skills in this area for letting the world know about our polar dinosaurs are invalu-able. Marion and Phil, with their talents and their long-term support, which is unpaid but definitely appreciated, are self-motivated volun-teers without whose help Dinosaur Dreaming would be much less known and not so well funded. Just like John Herman and the flying fox a decade before, they saw a problem and went about solving it on their own initiative. This kind of unasked-for help has made possible the ac-complishment of things we ourselves did not originally anticipate, and has made the overall project all the more productive and imaginative.

Notes

1. Except where explicitly used otherwise, the word "dinosaur" as used in this book has the conventional meaning. This is, it is shorthand for "dinosaurs other than birds." Strictly speaking, dinosaurs are birds and while the phrase "non-avian dinosaurs" could be used throughout and its employment would be strictly correct, it is more cumbersome than this single footnote.

2. Godthelp et al. (1992).

3. Woodburne & Case (1996).

4. In 1953, Mawson was a living legend in Australia. Shortly after the turn of the century, he went to Antarctica, where he carried out pioneering geological work and was a member of the first party to reach the south magnetic pole in 1909. Three years later, he led a three-man party which set out from Commonwealth Bay to explore inland. This trip, a 2,000-mile trek on foot (which he barely survived; neither of the others did), is recounted in his 1915 book, *Home of the Blizzard,* which is part of the lore of the Heroic Age of Antarctic exploration. Mawson returned to Australia and maintained his interest in Antarctic geological exploration. He also turned his energies to exploring South Australia geologically and thus was in a good position to assist Stirton as he did.

5. Andrews (1953).

6. Richard Tedford accompanied Stirton on a number of expeditions that explored the late Cenozoic of Australia in the 1950s and early 1960s. He subsequently returned on his own and continues to maintain an interest in this area. His other major research interests are the evolution of the Carnivora and mammalian biostratigraphy, particularly of North America and China. He has spent most of his professional career at the American Museum of Natural History, New York.

7. John Long is now Curator of Vertebrate Paleontology at the Western Australian Museum, Perth. There he continues his research on the evolution of fish, particular those of the Paleozoic. His work has resulted in the publication of a popular book on these fossils, *The Age of Fishes* (Long 1995). He has also published a popular book entitled *Dinosaurs of Australia and New Zealand* (Long 1998).

Tim Flannery is now Director of the South Australian Museum, Adelaide. He has carried out investigations of the terrestrial mammals of the southwest Pacific area and has written a number of books summarizing the mammalian fauna of that region. He has also written an informative and entertaining account of his travels there, *Throwim Way Leg* (Flannery 1998). He has edited a number of historical accounts of Australian exploration. The work for which Tim is best known to the Australian public is *The Future Eaters: An Ecological History of the Australasian Lands and Peoples* (Flannery 1994). A similar book about the United States, entitled *Conceived in Liberty,* is in press.

8. The holotype of *Agrosaurus macgillivrayi* was long thought to have been the first specimen from Australia to be recognized as a dinosaur and described as such in the scientific literature. But evidently that specimen is not from Australia at all, but from England. See Chapter 9 and Vickers-Rich, Rich, McNamara, & Milner (1999).

9. Ralph Molnar is now Curator of Vertebrate Paleontology at the Queensland Museum, Brisbane. His primary interests are the evolution of crocodiles and dino-

saurs. His principal research projects are describing and analyzing in detail two of Australia's most complete dinosaurs, the ankylosaur *Minmi* and the ornithopod *Muttaburrasaurus*.

10. The three scientific papers on the astragalus that have been published to date are Molnar, Flannery, & Rich (1981, 1985) and Welles (1983).

11. Warren et al. (1991)

12. Warren & Hutchinson (1983).

13. Dong (1985).

14. Shishkin (1991).

15. Gondwana is the collective name given to all the southern landmasses plus India. In the Middle Jurassic period, all of them were joined together after having split away from the northern landmasses (Laurasia). Prior to that, most of the landmasses of the Earth had been joined together in a single giant continent, Pangea. The Gondwana landmasses in turn began to separate from one another in the Late Jurassic period. Eastern Gondwana refers to Australia, New Zealand, and Antarctica, which were among the last of the Gondwana landmasses to separate from each other.

16. Mike Archer is now Director of the Australian Museum, Sydney. His principal research activity for the past two decades has been centered on late Cenozoic mammalian faunas of Queensland, with an emphasis on the Riversleigh district in the northwestern part of the state. His research program is summarized in Archer, Hand, & Godthelp (1991).

17. The explosive AN GELIGNITE 60 formerly manufactured by ICI Australia (now Orica) was quite similar in properties and effect to some of the well-known dynamite products of North American explosives manufacturers. Such explosives based on nitroglycerine are now virtually a thing of the past in many parts of the world; modern explosives offer safety, cost effectiveness, manufacturing convenience, and physiological benefits (no headaches brought about by fumes of nitroglycerine). (Alastair Blaikie, personal communication)

18. Rich et al. (1988).

19. Rich & Rich (1989).

20. For Leaellyn's perspective about this, see Rich (1995).

21. Wagstaff & McEwan-Mason (1989).

22. Idnurm (1985).

23. Douglas & Williams (1982).

24. Bob Gregory is now based at Southern Methodist University, Dallas, Texas. Along with colleagues also located there, he continues his interest in assessing the mean annual temperatures that prevailed in Australia during the Early Cretaceous.

25. Gregory et al. (1989) and Constantine, Chinsamy, Rich, & Vickers-Rich (1998).

26. Gross, Rich, & Vickers-Rich (1993).

27. Rich & Vickers-Rich (1994).

28. Douglas (1969, 1972).

29. Parrish et al. (1991).

30. Wegener (1915).

31. The symposium was held in New York on November 15, 1926. The papers presented there were published in van Waterschoot van der Gracht et al. (1928). Hallam (1973) is a readable history of the development of the ideas of continental drift and plate tectonics that was written shortly after the theory of plate tectonics had become well established; it thus has an immediacy about it.

32. If there can be said to be a seminal paper from which grew the geophysical revival of the idea that the continents move, it is Vine & Matthews (1966). In that paper, they pointed out how the pattern of magnetic anomalies on the ocean floor could be best explained by the idea that new crust was generated at the mid-ocean

ridges and moved out from there. The rapid growth of knowledge about the geology of the ocean floor that occurred in the decades after World War II set the stage for the birth of the idea of plate tectonics.

33. Du Toit (1937).

34. Luis Alvarez was, among other things, the senior author of the watershed (and by the way, eminently readable) paper which first proposed that the dinosaurs had become extinct after the impact of an asteroid. He based his theory on the abundance of the element iridium in the rock at the boundary between the Mesozoic and Cenozoic Eras (Alvarez, Alvarez, Asaro, and Michel 1980). He tells the story in his autobiography (Alvarez 1987) of how in 1939, although he had already developed a device that could have easily detected the soft neutron emissions predicted to accompany the nuclear fission process that had recently been discovered at the time, he failed to do so because he made only a brief attempt, not thinking the effort worthwhile. He spent an hour suitably modifying the device and five minutes trying to see the effect. He could have easily made the device a million times more sensitive for this experiment and had he done so, he would have detected the soft neutrons. In the meantime, Joliot, Kowarski, and von Halban in Paris and Fermi and Szilard at Columbia University spent months constructing suitable devices to do the same experiment because they were convinced the objective was worthwhile and eventually, after much effort, they did find the soft neutron emissions.

35. Metcalfe (1996).

36. Domack, Fairchild, & Anderson (1980).

37. Warren, Rich, & Vickers-Rich (1997).

38. Rich & Vickers-Rich (1994).

39. Dong & Azuma (1997).

40. Chinnery et al. (1998).

41. Flannery (1994).

42. Vickers-Rich & Rich (1993, 1999).

43. Lawson (1970).

44. Seeley (1891).

45. MacGillivray (1852).

46. Jukes (1847).

47. von Huene (1906).

48. Archer, Flannery, Ritchie, & Molnar (1985); Flannery, Archer, Rich, & Jones (1995).

49. Among the many books on fossil preparation and field techniques are Feldman, Chapman, & Hannibal (1989), Kummel & Raup (1965), Leiggi & May (1994), Rixon (1976), and Whitelaw & Kool (1991).

50. Before the jaw was found, the widely accepted view was that placentals appeared in Asia by the end of the Early Cretaceous (the same age as the jaw), marsupials evolved in early Late Cretaceous North America (their oldest record being slightly younger), and that monotremes had evolved in Australia by the late Early Cretaceous (Marshall & Kielan-Jaworowska 1992; Rich 1991). In this view, placentals had reached North America well before the end of the Cretaceous. Subsequently, at about the Cretaceous-Paleogene boundary, marsupials and placentals are thought to have entered South America, where they encountered monotremes that had probably crossed Antarctica from Australia, if they did not originate in South America (Pascual, Jaureguizar, & Prado 1996). This theory holds that both marsupials and placentals continued southward to reach the Antarctic Peninsula by the Eocene (Marenssi, Reguero, Santillana, & Vizcaino 1994), but from there only the marsupials reached Australia. Rats were then thought to have been the first terrestrial placentals to have reached Australia, arriving there in the Pliocene (Rich 1991).

51. Chamberlin (1890), p. 92.

52. The aptly titled *Tyranny of Distance* by the historian Geoffrey Blainey (1966) is a thought-provoking book about how the physical separation of Australia from other parts of the world has shaped the society of that continent economically, socially, and intellectually.

53. *Ausktribosphenos nyktos* was first described and named in Rich et al. (1997). A detailed critique of the tentative identification of *A. nyktos* as a placental appeared in Kielan-Jaworowska, Cifelli, & Luo (1998) to which a reply was made by Rich, Flannery, & Vickers-Rich (1998). An expanded description of *A. nyktos*, accompanied by the initial description of a second, quite different mammal from the Flat Rocks site, *Teinolophos trusleri,* appears in Rich et al. (1999).

54. Kielan-Jaworowska, Cifelli, & Luo (1998).

55. Don Russell's principal research interest is early Cenozoic mammals, both their systematics and paleobiogeography. In the latter area, in addition to doing much original work, he has published two important syntheses (Savage & Russell 1983 and Russell & Zhai 1987).

56. The Chinese mammal was named and described in Chow & Rich (1982). Wang, Clemens, Hu, & Li (1998) is an interesting follow-up to that paper. Wang and colleagues describe and analyze an upper molar found at the same site that probably belongs to the same genus.

57. A very readable account of the history of discovery and excavation of dinosaurs and mammals at Como Bluff during the nineteenth century is given in Ostrom & McIntosh (1966).

58. Anusuya Chinsamy has a combined appointment as a curator at the South African Museum and Associate Professor at the University of Cape Town. A semi-popular account of the methods by which she analyzes the microstructure of fossil bones is to be found in Chinsamy & Dodson (1995).

59. Constantine, Chinsamy, Rich, & Vickers-Rich (1998).

60. Russell, Beland, & McIntosh (1980).

61. Phil Currie is based at the Royal Tyrrell Museum of Palaeontology, Drumheller, Alberta, Canada, a museum which he has been with from the time of its founding. His principal research interest is theropod dinosaurs. He was a principal organizer and leader of the highly successful Canada-China Dinosaur Project, 1986–1991. He was the senior editor on the *Dinosaur Encyclopedia* (Currie & Padian 1997). He has also written several popular books (Currie 1991, Currie & Koppelhus 1996, and Currie, Koppelhus, & Sovak 1998).

62. Currie, Vickers-Rich, & Rich (1996).

63. Metcalfe (1996).

64. McKenna (1973).

65. McKenna & Bell (1997).

66. Stanhope et al. (1998b), p. 9969.

67. Waddell, Cao, Hauf, & Hasegawa (1999).

68. Stanhope et al. (1998a).

69. Clemens & Nelms (1993).

70. Rich, Gangloff, & Hammer (1997).

71. Rich, Gangloff, & Hammer (1997).

72. Alexander (1997), Etzioni (1999), Gardner (1993), Wilkinson (1997).

73. Power (1994).

74. Peters & Waterman (1982).

Literature Cited

Alexander, T. 1997. *Family Learning: The Foundation of Effective Education.* London: Demos. (57 pp.)

Alvarez, L. W. 1987. *Alvarez: Adventures of a Physicist.* New York: Basic Books.

Alvarez, L. W., W. Alvarez, F. Asaro, & H. V. Michel. 1980. Extraterrestrial cause for the Cretaceous-Tertiary extinction: Experimental results and theoretical interpretation. *Science* vol. 208, pp. 1095–1108.

Andrews, R. C. 1953. *All about Dinosaurs.* New York: Random House.

Archer, M., T. F. Flannery, A. Ritchie, & R. E. Molnar. 1985. First Mesozoic mammal from Australia—An early Cretaceous. *Nature,* vol. 318, pp. 363–366.

Archer, M., S. J. Hand, & H. Godthelp. 1991. *Riversleigh: The Story of Animals in Ancient Rainforests of Inland Australia.* Balgowlah: Reed Books.

Blainey, G. 1966. *The Tyranny of Distance: How Distance Shaped Australia's History.* Melbourne: Sun Books.

Chamberlin, T. C. 1890. The method of multiple working hypotheses. With this method the dangers of potential affection for a favorite theory can be circumvented. *Science* (old series) vol. 15, pp. 92–96. Reprinted in 1965: *Science,* vol. 148, pp. 754–759.

Chinnery, B. J., T. R. Lipka, J. I. Kirkland, J. M. Parrish, & M. K. Brett-Surman. 1998. Neoceratopsian teeth from the Lower to Middle Cretaceous of North America. In S. G. Lucas, J. I. Kirkland, & J. W. Estep (eds.), *Lower and Middle Cretaceous Terrestrial Ecosystems: New Mexico Museum of Natural History and Science Bulletin,* vol. 4, pp. 297–302.

Chinsamy, A., & P. Dodson. 1995. Inside a dinosaur bone. *American Scientist,* vol. 83, pp. 174–180.

Chow, M., & T. H. Rich. 1982. *Shuotherium dongi,* gen. et sp. nov., a therian with pseudotribosphenic molars from the Jurassic of Sichuan, China. *Australian Mammalogy,* vol. 5, pp. 127–142.

Clemens, W. A., & L. G. Nelms. 1993. Paleoecological implications of Alaskan terrestrial vertebrate fauna in latest Cretaceous time at high paleolatitudes. *Geology,* vol. 21, pp. 503–506.

Constantine, A., A. Chinsamy, T. H. Rich, & P. Vickers-Rich. 1998. Periglacial environments and polar dinosaurs. *South African Journal of Science,* vol. 94, pp. 137–141.

Currie, P. J. 1991. *The Flying Dinosaurs.* Red Deer, Alberta: Red Deer College Press.

Currie, P. J., & E. B. Koppelhus. 1996. *101 Questions about Dinosaurs.* New York: Dover Publications Inc.

Currie, P. J., E. B. Koppelhus, and J. Sovak. 1998. *A Moment in Time with* Centrosaurus. Calgary: Troodon Productions Inc.

Currie, P. J., & K. Padian (eds.). 1997. *Encyclopedia of Dinosaurs.* San Diego, Calif.: Academic Press.

Currie, P. J., P. Vickers-Rich, & T. H. Rich. 1996. Possible oviraptorosaur (Theropoda, Dinosauria) specimens from the Early Cretaceous Otway Group of Dinosaur Cove, Australia. *Alcheringa,* vol. 20, pp. 73–79.

Domack, E. M., W. W. Fairchild, & J. B. Anderson. 1980. Lower Cretaceous sediment from the East Antarctic continental shelf. *Nature,* vol. 287, pp. 625–626.

Dong, Z. 1985. A Middle Jurassic labyrinthodont (*Sinobrachyops placenticephalus* gen. et sp. nov.) from Dashanpu, Zigong, Sechuan Province. *Vertebrata Palasiatica,* vol. 23, pp. 301–307.

Dong, Z., & Y. Azuma. 1997. On a primitive neoceratopsian from the Early Cretaceous of China. In A. Dong (ed.), *Sino-Japanese Silk Road Dinosaur Expedition,* 68–89. Beijing: China Ocean Press.

Douglas, J. G. 1969. The Mesozoic Floras of Victoria, Parts 1 and 2. *Memoir of the Geological Survey of Victoria,* vol. 28.

Douglas, J. G. 1972. The Mesozoic Floras of Victoria, Part 3. *Memoir of the Geological Survey of Victoria,* vol. 29.

Douglas, J. G., & G. E. Williams. 1982. Southern polar forests: The Early Cretaceous floras of Victoria and their palaeoclimatic significance. *Palaeogeography, Palaeoclimatology, Palaeoecology,* vol. 39, pp. 171–185.

Du Toit, A. L. 1937. *Our Wandering Continents.* Edinburgh, Scotland: Oliver and Boyd.

Etzioni, A. 1999. *The Parenting Deficit.* London: Demos.

Feldman, R. M., R. E. Chapman, & J. T. Hannibal (eds.). 1989. *Paleotechniques.* University of Tennessee, Knoxville: Paleontological Society Special Publication No. 4.

Flannery, T. F. 1994. *The Future Eaters: An Ecological History of the Australasian Lands and Peoples.* Chatswood, New South Wales: Reed Books.

Flannery, T. F. 1998. *Throwim Way Leg: An Adventure.* Melbourne: Text Publishing Company.

Flannery, T. F., M. Archer, T. H. Rich, & R. Jones. 1995. A new family of monotremes from the Cretaceous of Australia. *Nature,* vol. 377, pp. 418–420.

Gardner, H. 1993. *Opening Minds.* London: Demos.

Godthelp, H., M. Archer, R. Cifelli, S. J. Hand, & C. F. Gilkeson. 1992. Earliest known Australian Tertiary mammal fauna. *Nature,* vol. 356, pp. 514–516.

Gregory, R. T., C. B. Douthitt, I. R. Duddy, P. V. Rich, & T. H. Rich. 1989. Oxygen isotopic composition of carbonate concretions from the lower Cretaceous of Victoria, Australia: Implications for the evolution of meteoric waters on the Australian continent in a paleopolar environment. *Earth and Planetary Science Letters,* vol. 92, pp. 27–42.

Gross, J. D., T. H. Rich, & P. Vickers-Rich. 1993. Dinosaur bone infection. *National Geographic Research and Exploration,* vol. 9, pp. 286–293.

Hallam, A. 1973. *A Revolution in the Earth Sciences: From Continental Drift to Plate Tectonics.* Oxford: Clarendon Press.

Hargreaves, D. 1994. *The Mosaic of Learning: Schools and Teachers for the Next Century.* London: Demos.

Idnurm, M. 1985. Late Mesozoic and Cenozoic palaeomagnetism of Australia—I: A redetermined apparent polar wander path. *Geophysical Journal of the Royal Astronomical Society,* vol. 83, pp. 399–418.

Jukes, J. B. 1847. *Narrative of the Surveying Voyage of the H.M.S. Fly.* London: Boone.

Kielan-Jaworowska, Z., A. W. Crompton, & F. A. Jenkins. 1987. The origin of egg-laying mammals. *Nature,* vol. 326, pp. 871–873.

Kielan-Jaworowska, Z., R. L. Cifelli, & Z. Luo. 1998. Alleged Cretaceous placental from down under. *Lethaia,* vol. 31, pp. 267–268.

Kummel, B., & D. Raup. 1965. *Handbook of Paleontological Techniques.* San Francisco: W. H. Freeman and Company.

Lawson, H. 1970. *While the Billy Boils.* Hawthorne, Victoria: Lloyd O'Neil.

Leiggi, P., & P. May. 1994. *Vertebrate Paleontological Techniques.* Vol. 1. Cambridge: Cambridge University Press.

Long, J. A. 1995. *The Age of Fishes: 500 Million Years of Evolution.* Sydney: University of New South Wales Press.

Long, J. A. 1998. *Dinosaurs of Australia and New Zealand and Other Animals of the Mesozoic Era.* Sydney: University of New South Wales Press.

MacGillivray, J. 1852. *Narrative of the Voyage of the H.M.S. Rattlesnake.* London: Boone.

McKenna, M. C. 1973. Sweepstakes, filters, corridors, Noah's Arks, and Beached Viking Funeral Ships in paleogeography. In D. H. Tarling & S. K. Runcorn (eds.), *Implications of Continental Drift to the Earth Sciences,* 1:295–308. London: Academic Press.

McKenna, M. C., & S. K. Bell. 1997. *Classification of Mammals above the Species Level.* Columbia University Press, New York.

Marenssi, S. A., M. A. Reguero, S. N. Santillana, & S. F. Vizcaino. 1994. Eocene land mammals from Seymour Island, Antarctica: Palaeobiogeographical implications. *Antarctic Science,* vol. 6, pp. 3–15.

Marshall, L. G., & Z. Kielan-Jaworowska. 1992. Relationships of the dog-like marsupials, deltatheroidans and early tribosphenic mammals. *Lethaia,* vol. 25, pp. 361–374.

Mawson, D. 1915. *The Home of the Blizzard.* London: William Heineman.

Metcalfe, I. 1996. Gondwanaland dispersion, Asian accretion and evolution of eastern Tethys. *Australian Journal of Earth Science,* vol. 43, pp. 605–623.

Molnar, R. E., T. F. Flannery, & T. H. Rich. 1981. An allosaurid dinosaur from the Cretaceous of Victoria, Australia. *Alcheringa,* vol. 5, pp. 141–146.

Molnar, R. E., T. F. Flannery, & T. H. Rich. 1985. Aussie *Allosaurus* after all. *Journal of Paleontology,* vol. 59, pp. 1511–1513.

Novas, F. E. 1997. Herrerasauridae. In P. J. Currie & K. Padian (eds.), *Encyclopedia of Dinosaurs.* San Diego: Academic Press, pp. 303–311.

Ostrom, J. H., & J. S. McIntosh. 1966. *Marsh's Dinosaurs: The Collections from Como Bluff.* New Haven, Conn.: Yale University Press.

Parrish, J. T., R. A. Spicer, J. G. Douglas, T. H. Rich, & P. Vickers-Rich. 1991. Continental climate near the Albian South Pole and comparison with climate near the North Pole. *Geol. Soc. Amer., Abstracts with Programs,* vol. 23, p. A302.

Pascual, R., E. O. Jaureguizar, & J. L. Prado. 1996. Land mammals: Paradigm for Cenozoic South American geobiotic evolution. In G. Arratia (ed.), *Contributions of Southern South America to Vertebrate Paleontology. Münchner geowissenschaftliche Abhandlungen: Reihe A, Geologie und Paläontologie,* vol. 30. München: Verlag Dr. Friedrich Pfeil, pp. 265–319.

Paul, G. S. 1988. *Predatory Dinosaurs of the World: A Complete Illustrated Guide.* New York: Simon & Schuster.

Paul, G. S. 1997. Dinosaur Models: The Good, the Bad, and Using Them to Estimate the Mass of Dinosaurs. In D. L. Wolberg, E. Stump, & G. D. Rosenberg (eds.), *Dino Fest International Symposium Volume,* pp. 127–154.

Peters, T. J., & R. H. Waterman. 1982. *In Search of Excellence.* New York: Harper & Row.

Power, M. 1994. *The Audit Explosion.* London: Demos.

Rich, L. 1995. My little dino. *Ranger Rick,* vol. 29, no. 2, pp. 32–39.

Rich, P. V., T. H. Rich, B. Wagstaff, J. McEwen Mason, R. T. Douthitt, R. T. Gregory, & A. Felton. 1988. Evidence for low temperatures and biologic diversity in Cretaceous high latitudes of Australia. *Science,* vol. 242, pp. 1403–1406.

Rich, T. H. 1991. Monotremes, placentals, and marsupials: Their record in Australia and its biases. In P. Vickers-Rich, J. M. Monaghan, R. F. Baird, & T. H. Rich (eds.), *Vertebrate Palaeontology of Australasia,* pp. 893–1070. Melbourne: Pioneer Design Studio in cooperation with the Monash University Publications Committee.

Rich, T. H. 2000. Australia: Vertebrate Paleontology. In R. Singer (ed.), *Encyclopedia of Paleontology.* Vol. 1. Chicago: Dearborn, pp. 140–149.

Rich, T. H., T. F. Flannery, & P. Vickers-Rich. 1998. Alleged Cretaceous placental from down under: Reply. *Lethaia,* vol. 31, pp. 346–348.

Rich, T. H., R. A. Gangloff, & W. Hammer. 1997. Polar Dinosaurs. In P. J. Currie & K. Padian (eds.), *Encyclopedia of Dinosaurs,* pp. 562–573. Academic Press.

Rich, T. H., & P. V. Rich. 1989. Polar dinosaurs and biotas of the Early Cretaceous of southeastern Australia. *National Geographic Research,* vol. 5, pp. 15–53.

Rich, T. H., & P. Vickers-Rich. 1994. Neoceratopsians & ornithomimosaurs: Dinosaurs of Gondwana origin? *National Geographic Research and Exploration,* vol. 10, pp. 129–131.

Rich, T. H., P. Vickers-Rich, A. Constantine, T. F. Flannery, L. Kool, & N. van Klaveren. 1997. A tribosphenic mammal from the Mesozoic of Australia. *Science,* vol. 278, pp. 1438–1442.

Rich, T. H., P. Vickers-Rich, A. Constantine, T. F. Flannery, L. Kool, & N. van Klaveren. 1999. Early Cretaceous mammals from Flat Rocks, Victoria, Australia. *Records of the Queen Victoria Museum,* 106.

Rixon, A. E. 1976. *Fossil Animal Remains: Their Preparation and Conservation.* London: Athelone Press.

Russell, D. E., P. Beland, & J. S. McIntosh. 1980. Paleoecology of the dinosaurs of Tendaguru (Tanzania). *Mémoires de la Société géologique de France, Paris,* mem. 59, no. 139, pp. 169–176.

Russell, D. E., & R-J. Zhai. 1987. *The Paleogene of Asia: Mammals and stratigraphy.* Paris: Mémoires du Muséum National d'Histoire Naturelle, Série C, Sciences de la Terre, Tome 52.

Savage, D. E., & D. E. Russell. 1983. *Mammalian Paleofaunas of the World.* Reading, Mass.: Addison-Wesley.

Schmidt, P. W., & B. J. J. Embleton. 1981. Magnetic overprinting in southeastern Australia and the thermal history of its rifted margin. *Journal of Geophysical Research,* vol. 86, pp. 3998–4008.

Seeley, H. G. 1891. On *Agrosaurus macgillivrayi* (Seeley), a Saurischian Reptile from the N. E. Coast of Australia. *Quarterly Journal of the Geological Society,* vol. 47, pp. 164–165.

Shishkin, M. A. 1991. A Late Jurassic labyrinthodont from Mongolia. *Paleontological Journal,* vol. 1, pp. 78–91 [English translation].

Stanhope, M. J., O. Madsen, V. G. Waddell, G. C. Cleven, W. W. de Jong, & M. S. Springer. 1998a. Highly congruent molecular support for a diverse superordinal clade of endemic African mammals. *Molecular Phylogenetics and Evolution,* vol. 9, pp. 501–508.

Stanhope, M. J., V. G. Waddell, O. Madsen, W. de Jong, S. B. Hedges, G. C. Cleven, D. Kao, & M. S. Springer. 1998b. Molecular evidence for multiple origins of Insectivora and for a new order of endemic African Insectivore mammals. *Proceedings of the National Academy of Scientists of the United States of America,* vol. 95, pp. 9967–9972.

van der Gracht, W. A. J. M. van Waterschoot, et al. (eds.). 1928. *Theory of Continental Drift: A Symposium.* Tulsa, Okla.: American Association of Petroleum Geologists.

Vandenberghe, J. 1988. Cryoturbations. In M. J. Clark (ed.), *Advances in Periglacial Geomorphology*, pp. 179–198. Chichester, U.K.: John Wiley and Sons, Ltd.

Vickers-Rich, P., & T. H. Rich. 1993. *Wildlife of Gondwana*. Sydney: Reed Books.

Vickers-Rich, P., & T. H. Rich. 1999. *Wildlife of Gondwana: Dinosaurs and Other Vertebrates from the Ancient Supercontinent*. Bloomington: Indiana University Press.

Vickers-Rich, P., T. H. Rich, G. McNamara, & A. Milner. 1999. Is *Agrosaurus macgillivrayi* Australia's oldest dinosaur? *Records of the Western Australian Museum Supplement*, vol. 57, pp. 191–200.

Vine, F. J., & D. H. Matthews. 1963. Magnetic anomalies over ocean ridges. *Nature*, vol. 199, pp. 947–949.

von Huene, F. (1906). Ueber die Dinosaurier der aussereuropäischen Trias. *Geologische und Palaeontologische Abhandlungen*. Berlin: Jena (N. F.) vol. 8, no. 12, pp. 97–156.

Waddell, P. J., Y. Cao, J. Hauf, & M. Hasegawa. 1999. Using novel phylogenetic methods to evaluate mammalian mtDNA, including amino acid-invariant sites-LogDet plus site stripping, to detect internal conflicts in the data, with special reference to the positions of hedgehog, armadillo, and elephant. *Systematic Biology*, vol. 48, pp. 31–53.

Wagstaff, B. E., & J. McEwen-Mason. 1989. Palynological dating of lower Cretaceous coastal vertebrate localities, Victoria, Australia. *National Geographic Research*, vol. 5, pp. 54–63.

Wang, Y., W. A. Clemens, Y. Hu, & C. Li. 1998. A probable pseudo-tribosphenic upper molar from the late Jurassic of China and the early radiation of the Holotheria. *Journal of Vertebrate Paleontology*, vol. 18, pp. 777–787.

Warren, A. A., & M. N. Hutchinson. 1983. The last labyrinthodont? A new brachyopoid (Amphibia, Temnospondyli) from the Early Jurassic Evergreen Formation of Queensland, Australia. *Philosophical Transactions of the Royal Society*, Series B, vol. 303, pp. 1–62.

Warren, A. A., L. Kool, M. Cleeland, T. H. Rich, & P. V. Rich. 1991. An Early Cretaceous labyrinthodont. *Alcheringa*, vol. 15, pp. 327–332.

Warren, A., T. H. Rich, & P. Vickers-Rich. 1997. The last last labyrinthodont? *Palaeontographica Abt. A*, vol. 247, pp. 1–24.

Wegener, A. 1915. *Die Entstehung der Kontinents und Ozeane*. Braunschweig: Vieweg. 1966 English translation by John Biram of the 4th German edition of 1929: *Origin of Continents and Oceans*. New York: Dover.

Welles, S. P. 1983. *Allosaurus* (Saurischia, Theropoda) not yet in Australia. *Journal of Paleontology*, vol. 57, p. 196.

Whitelaw, M., & L. Kool. 1991. Techniques used in preparation of terrestrial vertebrates. In P. Vickers-Rich, J. M. Monaghan, R. F. Baird, & T. H. Rich (eds.), *Vertebrate Palaeontology of Australasia*, pp. 173–200. Melbourne: Pioneer Design Studio in cooperation with the Monash University Publications Committee.

Wilkinson, H. 1997. *Time Out: The Costs and Benefits of Paid Parental Leave*. London: Demos.

Woodburne, M. O., & J. A. Case. 1996. Dispersal, vicariance, and the Late Cretaceous to early Tertiary land mammal biogeography from South America to Australia. *Journal of Mammalian Evolution*, vol. 3, pp. 121–161.

Index

THOMAS H. RICH is Curator of Vertebrate Paleontology at Museum Victoria in Melbourne and co-author (with Patricia Vickers-Rich) of *Wildlife of Gondwana: Dinosaurs and Other Vertebrates from the Ancient Supercontinent*, also published by Indiana University Press.

PATRICIA VICKERS-RICH holds a Chair in Paleontology at Monash University, where she lectures in the Earth Sciences department. She is also Director of the Monash Science Centre.